Wolff's Guide to the London Metal Exchange

by Rudolf Wolff & Co Ltd

5th Edition 1995

Published by:

Metal Bulletin Books Ltd

Park House
Park Terrace
Worcester Park
Surrey KT4 7HY
England

220 Fifth Avenue
New York
NY 10001

16 Lower Marsh
London SE1 7RJ
England

ISBN: 0-947671-92-7

ISSN: 0144-5960

British Library Cataloguing in Publication Data

Wolff's Guide to the London Metal Exchange - 5th edition

1. London Metal Exchange

I. Rudolf Wolff & Co

332.64′4 HD9 521.9.L6

ISBN: 0-947671-92-7

ISSN: 0144-5960

Printed and bound in Great Britain by
Redwood Press Ltd
Trowbridge, Wiltshire

Wolff's Guide to the London Metal Exchange

5th Edition 1995

NULLUS IN VERBA

Rudolf Wolff & Co Ltd

Rudolf Wolff & Co was founded in 1866, at a time when London was becoming the major clearing house for the world's surplus metals. Gradually a metals 'futures' market evolved and in 1877 Rudolf Wolff was amongst the group of merchants who formed the London Metal Exchange.

Over the years Rudolf Wolff has become one of the leading brokers and Ring Dealing Members of the London Metal Exchange and members of the Company have played an active role in the organisation and running of the Exchange.

In 1971, Noranda Inc of Canada acquired a 51 % interest in Rudolf Wolff & Co Ltd which was increased to 100% in 1982. Noranda is one of the world's major resources companies with extensive interests in mining, metallurgy, forestry and energy.

Although Rudolf Wolff's activities have traditionally been centred on the London Metal Exchange, it has expanded to encompass the full range of futures and options in soft commodities, energy and financial futures as well as bullion and foreign exchange. The Group trades on all the world's leading commodity and financial futures exchanges and through its international network, it is ideally placed to serve the increasingly diverse requirements of its international clientele.

Rudolf Wolff is regulated in the conduct of its investment business by the Securities and Futures Authority (SFA).

Rudolf Wolff & Co Ltd is a member of the following exchanges:

LONDON

London Bullion Market Association

London Futures & Options Exchange
(Coffee, Cocoa, Sugar, Freight)

International Petroleum Exchange of London

London International Financial Futures Exchange

Baltic International Freight Futures Exchange

NEW YORK

Commodity Exchange Inc

Rudolf Wolff & Co Offices

LONDON

Rudolf Wolff & Co Ltd, Plantation House, 31-35 Fenchurch Street,
London EC3M 3DX
Telephone: (44) 171-626 8765. Telex: 885034. Fax: (44) 171-626 3939.

NEW YORK

Rudolf Wolff & Co Inc, 295 Madison Avenue, New York, NY 10017,
USA
Telephone: (1) 212-370 5115. Telex: 229386. Fax: (1) 212-532 3975.

TOKYO

Rudolf Wolff KK, Sanbancho MS Building, 7th Floor 20 Sanbancho,
Chiyoda-ku, Tokyo, Japan.
Telephone: (81) 3-3222 0171. Telex: J29695. Fax: (81) 3-3222 0175.

HAMBURG

Rudolf Wolff & Co GmbH, Rathaushof, Klein Johannisstrasse 2-4,
20457 Hamburg 11, Germany.
Telephone: (49) 40-361041. Telex: 215661. Fax: (49) 40-372144.

BUDAPEST

Rudolf Wolff & Co KFT, Hermina Irodanaz, Hermina ut 17, 11th Floor,
XIV Budapest, Hungary.
Telephone: (36) 1-268 0674. Fax: (36) 1-268 0677.

WARSAW

Rudolf Wolff & Co SP.Z.O.O., Ul. Olesinska 21 R 115, PL-02-548,
Warsaw, Poland.
Telephone: (48) 22-450 999. Fax: (48) 22-454 206.

also Agents in

Melbourne, Australia; Johannesburg, South Africa; Santiago, Chile;
Mexico City, Mexico; Seoul, South Korea.

The Metal Bulletin Group

Publications and Services

In addition to serving the metals industry, Metal Bulletin's information activities span the worlds of mining, non-metallic minerals, futures and options, and wool and textiles, through a range of magazines, surveys, directories, reference books and conferences.
These include 'Trading in Metals' which, together with 'Wolff's Guide to the London Metal Exchange', provides an unrivalled insight into physical and terminal market metal trading.

A brief selection of other titles and services appears below:

Journals

Metal Bulletin
MBM – Metal Bulletin Monthly
Futures & Options World
Futures & Options Plus
Clearing & Settlement
Derivatives in Fund Management

Books

Metal Traders of the World
Non-Ferrous Metal Works of the World
Philipp Brothers – The Rise and Fall of a Trading Giant
The Tin Men
Ferro-Alloy Directory and Databook
International Scrap Directory
World Aluminium – a Metal Bulletin Databook
World Copper Databook
Metal Bulletin's Prices & Data
International Futures & Options Databook
The World's Clearing Houses

Information Services

Base Metals Monthly
Ferro-Alloys Monthly
Fast Track
Prices Service
Metal Bulletin's Prices & Commentary
MetalsFinance

Metal Bulletin plc offices

Park House, Park Terrace, Worcester Park, Surrey KT4 7HY, England
Tel: +44 171 827 9977. Fax: +44 181 337 8943. Telex: 21383.

16 Lower Marsh, London SEI 7RJ, England
Tel: +44 171 827 9977. Fax: +44 171 928 6539. Telex 917706.

Metal Bulletin Inc

220 Fifth Avenue, New York, NY 10001, USA
Tel: +1 212 213 6202. Fax: +1 212 213 6619.

Contents

Continued

Appendices

Illustrations

LME Members in 1898 – already a substantial group

Re-opening Tin on LME in 1949

The LME ring at Plantation House in 1992

The technology revolution! A trading station at Rudolf Wolff's London office in 1995

The new LME ring on Leadenhall Street

The Rudolf Wolff dealing team on the new LME ring

LME ring at Whittington Avenue in 1973

Rudolf Wolff dealing room in 1971

The old timers – a typical grouping of Canadian miners in the early years

The traditional mine/smelter complex – the Horne at Noranda, Quebec

Aluminium – a versatile metal. Wheels for the automobile industry, extrusions for structural applications and compact discs for the home entertainment and information technology markets. (By courtesy of Dubai Aluminium Company Limited).

Copper anodes being loaded into the CCR refinery in Montreal

The modern miner with his equipment

A typical modern open pit mine in British Columbia

Rudolf Wolff, a founder member of the Exchange in 1877

James Wolff, Chairman of the LME Board 1946-1961

F F (Freddie) Wolff, Chairman of the LME Committee 1970-1977 and member of the LME Board

John Wolff, Chairman of the Board of the LME 1990-1992

Foreword by Bill Harker, Managing Director, Rudolf Wolff & Co Ltd

We tend to dwell on anniversaries in introducing new editions of Wolff's Guide to the London Metal Exchange. Ten years ago October 1985 marked the beginning of what we are all unhappy to remember as "the tin crisis" when the future of the London Metal Exchange seemed bleak indeed. Happily the subsequent ten years has included more growth, more change and more opportunity than probably any other decade in the 130 year history which we at Rudolf Wolff, now the unique survivor amongst the LME founder members, are proud to have shared with the Exchange

Ten years ago the LME members transacted futures contracts entirely as principals both to clients and other members, typically only up to three months forward, were lightly supervised by the Bank of England, and, other than in copper, served mostly the merchant community. In the intervening period we have seen the successful establishment of inter-member clearing by the London Clearing House, increasingly detailed governmental regulation, not only by the Securities and Futures Authority, our appointed regulator, but also by US and EC governmental bodies, the arrival of traded options contracts, pioneered by Rudolf Wolff, the 27 month futures contract with revised quality specifications in most metals to meet evolving industry standards, and, above all, an explosion in usage by professional risk managers throughout the metals industries and by professional trading managers in the investment community. The pace of development has been such that, with painful memories dulled by time, even tin trading has regained acceptance.

Continued

With so much that is new to adapt to, the true challenge for the LME has been to preserve what is best in its old tradition. We have been able to combine cleared contracts and regulation with 24 hour market-making so that we are still the only futures exchange in the world whose members can satisfy their clients' trading needs with an exchange-regulated contract at any hour of the day or night. Proponents of new hedging techniques have used the price discovery mechanism of the LME to invent new instruments governing complex counterparty relationships with their customers, but the professional hedging community still presses for the extension of LME contracts because they value their fungibility and the brokerage traditions of the members who execute them. Perhaps this ability to combine the old with the new is the main reason why cleared volume on the exchange has risen from an average of 224,000 lots per month in 1986 to 3,974,000 in 1994.

Wolff's Guide is now one of these valued traditions and for the 5th edition we have asked an old and trusted friend, Trevor Tarring, to help us make the transition from the previous format to one designed to address the changing role of the exchange as it points itself towards the 21st century. Trevor has long been a cornerstone of the Metal Bulletin group and a forceful spokesman for the LME in the metals industry. He has taken on the task of writing the new edition as a 'user friendly' tool for practical hedgers, investors and traders, focusing on the interests of you the reader rather than us the market technicians. I would like to thank him and all the others responsible for the book's contents, in particular David Waite, who spent many hours translating the complicated workings of our market into a series of understandable concepts, and the LME for allowing us to reprint those parts of the exchange's Rules which appear in the Appendices. We all hope that you the reader continue to find it worthwhile.

Acknowledgements
With acknowledgements to the following contributors and helpers:
Bill Harker, David Waite, Brian Woodbridge, William Adams and Michael Davis at Rudolf Wolff; Trevor Tarring, David King and the LME staff.

CHAPTER 1

Setting the scene

Pick up a local newspaper almost anywhere in the developed world and the chances are you will read an account of a horrifying road accident in which one or several people are killed. Yet in the same edition, quite likely, there will be a feature on the new season's cars and how they are reliable, economical, faster, more sporty and generally the stuff that fantasies are made of. Somewhere between these two extremes, nobody seems to have recorded the thousands of day-to-day journeys made by thousands of people over hundreds of thousands of miles that were strictly normal and functional. People got to work, to school, to relatives, to appointments by car. If they had gone any other way, it would have been slower, harder or impossible.

The futures industry, of which the London Metal Exchange is part, suffers from the same kind of one-sided reporting. The only time it hits the headlines in all the papers is when someone – usually a big corporation – has lost millions of dollars on futures trading. As with the reporting on the car accidents, the stories are long on horrifying details, short on analysis of the causes. Often the cause turns out to be the same as the cause of the car accidents – someone has been irresponsible, criminally negligent or, at the least, out of their depth.

Similarly in futures, there is almost never any reporting of the thousands of people who routinely enter into millions of dollars' worth of transactions every day. They not only come to no harm by so doing, they are actually avoiding harm. And, as with the car drivers, they are getting to where they want to be more quickly and cost-effectively than if they had gone any other way. Often the use of the futures markets is the only way.

The contrast between these two pictures of the futures business is great enough. But on top of this, far fewer people have practical experience of the use of futures markets – either directly or even by second-hand experience – than have experience of motoring. So it is perhaps not surprising that the myths about the futures industry being a gambling den persist, even though the futures markets have been with us a good deal longer than the motor car.

It is the purpose of this book to show how one futures market, the London Metal Exchange, is a useful and normal tool for use by those exposed to the risk of metal price movements. They may be miners, smelters, semi-fabricators, end users or scrap merchants; they may be banks, investors, government agencies or funds. They all routinely use the Exchange. They may occasionally have the odd minor bump or scrape, just as one does in today's urban motoring. But they have the skill and understanding of the tool they are using to avoid a major accident.

Accidents, we have said, can be due to three kinds of human failing – irresponsibility, criminal negligence or being out of one's depth. The last of these failings is the one this book is primarily dedicated to eliminating. Most people forget very quickly the agony of learning to drive, and thereafter have difficulty relating to the agonies of those who have not yet learned (witness any parent trying to teach their child to drive).

In the same way, operations on futures markets appear, to the uninitiated, to be fiendishly complicated and fraught with danger. Whereas to practitioners and their experienced clients it appears the most natural thing in the world to interpret each new form of price risk that arises in the physical market in terms of the ways to offset it by the use of the futures market. It would be too much to promise that reading this book will, on its own, enable any newcomer to the metals field to join the experts in the use of the Metal Exchange. But it does undertake to attempt to carry the reader forward from a position of total ignorance of the use of the futures markets to a degree of familiarity from which he or she can, by their own endeavours, progress to expertise.

SUPERVISING THE OPERATORS

One aspect of the educational nature of this book that has not been forgotten is the question of supervision of trading personnel. If a reasoned analysis is, exceptionally, made of a futures-related disaster, it often turns out that the level at which the necessary skills were lacking was not the front line operators, but the management responsible for supervising them.

To someone who has already progressed up the corporate ladder, it may seem distasteful to have to admit to a lesser level of skill in the use of the futures markets than that of the person one is supervising. But the effects of failing to admit that fact, and of then failing to rectify it, could result in the manager featuring in a headline about a futures disaster.

Similarly, managers in, for example, banking concerned with financing metals production, trading or fabrication, may be familiar with the use of futures in other fields such as exchange rates or stock indices. However, there are subtle but vital differences between operations on the London Metal Exchange and those on the financial futures markets. Failure to recognise these differences could result in a dangerous misunderstanding of the hedge programme used by a mine or smelter whose financial health is as important to the bank as it is to the company itself.

But the most important lesson of all that this book seeks to impart is that exposure to metal price risk means that danger is already present. It is to offset that danger that experienced metal position managers make skilled use of the futures market. To refuse to use the futures market because the operations are unfamiliar and appear dangerous is not the way to play it safe. It is merely a way to stay with the primary metal price risk, and to seek to ignore that risk because it is familiar.

There is a further dimension to a sound understanding of the nature of the London Metal Exchange, and this comes into play even if one has no need to make use of the Exchange. Most transactions in the seven metals quoted on the LME are governed to a greater or less degree by the quotations on the Exchange. So an intelligent understanding of the trading operations that produce the LME's closing prices is a definite advantage compared with simply sitting by the telephone wondering what prices a malign Fate will hand down today.

Back in a more innocent age of advertising, a soap powder manufacturer used the slogan "A million housewives can't be wrong". The same can be said of the hundreds of Exchange clients that use its contracts day in, day out as an integral part of their risk management process.

CHAPTER 2

Making a start

This chapter presupposes that the reader has accepted the statement of metal price risk in the opening chapter, and wishes, at least in principle, to hedge that risk. If, however, the reader does not accept the explanation of metal price risk set out there, and still considers operations on a futures market to be gambling, he should at least undertake a little back testing. That is to say, he should look at actual transactions he has undertaken in both rising and falling markets in the past where there was a time interval between entering and leaving a position of metal price risk.

We will take the simplest case of a smelter or refiner, producing metal in unwrought form and getting paid for it on a formula geared to the LME price. It is pretty clear what his back testing will tell him. In a rising market he makes windfall profits (on which he will almost certainly be taxed) and in a falling one he suffers a reduction in profitability over his production costs. Eventually, prices will fall sufficiently far that his profits become non-existent, and then losses. The back testing this producer should undertake will be understood by anybody who has used a business planning software programme – he should ask himself "What if?". What if he had hedged his metal sale as soon as the metal was poured and had started to enter the phase of processing and shipment?

If the back testing is conducted completely objectively, it will show, more or less precisely, that hedging with futures would have moderated the windfall profit in a rising market. More importantly, it would also have moderated the reduced profitability and losses associated with a falling market. When it comes to long-term survival, minimising the period of exposure to loss is more important than the occasional windfall profit.

DON'T FORGET OPTIONS

Perhaps the most important item of all to be back-tested is the use of options as well as futures as the tool to achieve the desired hedge-protection. Of all the forms of derivatives, options are the least understood and the most maligned. Yet in application, their value as a tool to be used in hedging is often easier to understand than that of futures. This is because an option acts like a ratchet. For example, it affords the option-holder a defence against a falling market, if that is what he fears; yet at the same time, it allows him to participate in a rising market.

The nature and use of options will be explained in more detail later in the book, but at this point, a few basic facts are essential:

There are two kinds of options – options on physical and options on futures. Of these, only options on futures can be exchange-traded, and not all options on futures are necessarily traded within the regulatory envelope of a futures market. Almost all the references in this book to options are to exchange-traded options on futures, and any reference to any other sort of option will be flagged as such.

Options cut two ways called, respectively, put options and call options. A PUT option gives the holder the right, but not the obligation, to SELL metal at a defined price at a specific date in the future to the person who sold him the option ("the grantor"). A CALL option gives the holder of the option the right but not the obligation to BUY metal from the grantor at a defined price at a defined date in the future. An option is created when the grantor sells the option to the option buyer for a consideration called the premium.

Because the premium in turn is a fluctuating figure, influenced by the strength or weakness of the demand for a given kind of option, it is much harder to give concrete examples of a hedge strategy implemented through options than it is of a strategy implemented through futures. For this reason, most futures textbooks tend to regard hedging with futures as degree-level stuff, but hedging with options as the province of PhD's! In this book we shall try to bridge this gap and to illustrate the use of futures and options even-handedly, as appropriate.

For the time being, however, we must leave the sceptical reader to undertake his back-testing before re-joining us. For it is absolutely vital that hedging should not be undertaken by those that secretly do not believe in it.

This is also a good point at which to make clear that hedging is not some kind of magic wand that can wave away all metal price problems. If the markets are in a secular downswing, for example, for three, four or even five years, it is unlikely that any hedging strategy will yield prices consistently 20, 30 or 50% above the prevailing market, much as our abovementioned miner would like it to do so. But locking in this year's price for next year's business by hedging is entirely practicable, throughout the downswing. And it has clearly beneficial effects on the average price received compared with doing no hedging at all. Again, the same is true for the semi-fabricator viewing price movements through the other end of the telescope. And for both parties, both futures and options can be used as appropriate to their risk management strategy.

KEEPING A BALANCED BOOK

But what about, say, a copper custom smelter who both buys his metal content and sells his refined metal on the basis of the LME price? Surely he can keep his metal book sufficiently well balanced that he is not exposed to metal price risk? In theory, this might be possible. In practice it isn't. Few people confront as many variables in their day to day operations as a custom smelter. To start with, he is buying metal in concentrates, in scrap or in residues. Even with the most reliable supplier and the most experienced operation of the smelter and/or refinery, metal outturn is always liable to differ from the theoretical. If the smelter expects to produce 600 tonnes this week and actually manages to sell those 600 tonnes back-to-back, it can easily happen that the outturn is only 500 tonnes. It's not funny being out of position on 100 tonnes of metal, especially if the positional imbalance is by being short and the market is rising. Equally likely, the production expectation is 600 tonnes and indeed 600 tonnes is produced, but the customers won't oblige with a total offtake between them on their long-term and spot purchases of 600 tonnes.

It's true the smelter can go to the LME (in trading hours at his broker) and buy or sell physical in the market to balance the books. But this is not an efficient way to use the exchange, and any smelter that consistently operated this way would be killed by the costs of putting metal on warrant and taking it off. In any case, he would be quite unable to move metal into and out of warehouse on a same-day basis, no matter how near he is to an LME-approved warehouse.

The same principles, of course, also apply to a semi-fabricator, with some differences of practice. He may, for example, be buying his raw metal as refined and selling the metal content of his semis on the basis of the refined price. But it is not at all certain that his customers will be sensitive to his desire to price the metal content of his output on the same LME price basis as applied to his purchases of raw metal. They may have their own pressing agenda which says that they have to buy the sheet, cable or tube at a fixed price for delivery a number of weeks or months forward. And if the semi-fabricator decides that his only safe course of action is to produce the semis first and then price them as spot sales, he is running a metal position throughout the process of manufacture, and it will be a long position. In a falling market, this is very dangerous. In any case, this fabricator will miss out on all the business tendered for at fixed prices, as competitors adopting the correct hedged approach to such quotations will be complying with the tender terms; a "make it then price it" fabricator will be unable to do this safely.

Even an international trader, who is not physically transforming his metal in any way, but merely shipping it from A to B, commonly faces metal price risk for the period the metal is on the water. Of course it is, theoretically, possible to get sale and purchase contracts so perfectly back to back that they are for the same date. But achieving this is rare. And in any case, the days when a profitable deal could be done for no more service than shipping the metal from A to B have long gone. Producer and consumer do not need a trader to arrange this for them.

NOTHING TO DO WITH END USERS?

Last but not least, we must consider the end-user of metal, producing capital goods for the electrical industry, for example, or automobiles, houses, white goods or factory buildings. All these people are large users of one or other of the base metals in one form or another. In metal industry terms, such businesses are often called second-line consumers as they buy their material from the first-line consumers, the semi-fabricators. Unfortunately, this position often leads these consumers to believe that they are not really part of the metal business, but part of the engineering or construction industry. From here it is only a short step to believing that metal price instability is something inherently outside their business experience, and therefore something they can do nothing about. This has been shown time and time again to be a most dangerous misconception.

It is clear how a miner, the first person in the metal industry chain, is squeezed between cost factors unrelated to the metal markets and a revenue stream entirely determined by those markets. But equally the end user, as the last person in the chain, is squeezed between cost inputs largely influenced by metal price movements and a market for his products which is, justifiably, indifferent to those price movements. Obviously, the question of volume of metal comes into it. A producer of, say, tiny electric motors for tape recorders or automotive accessories, may use only one or two tonnes a month of copper in his windings. Such a volume is, indeed, below the horizon of what is practically hedgeable. But a producer of motors for washing machines, or heat exchangers for refrigerators, may well have a sufficient monthly or quarterly throughput of copper or aluminium to make it worth his while to hedge that position.

So, however we look at it, dealing with physical metal normally gives rise to positional risk, and positional risk gives rise to price risk; the longer the position is open, the greater the risk. So to anybody in the base metals industry, it is comforting to know that the LME is a prime example of what is called a trade-oriented futures exchange. In other words, the Exchange's modus operandi is structured to facilitate its use by trade clients. In this respect it differs from many other commodity futures exchanges around the world, including most of the other metal futures exchanges.

It might appear from this, therefore, that use of the exchange by non-trade clients – banks, funds and affluent speculators – would be disadvantaged to the point where it would happen only rarely or not at all. However, this is far from the case, and recent years have given more convincing and practical proof of this than for some time past. Indeed, it is understood by those who study the operations of the Exchange objectively that non-trade users are vital to the smooth operation of the market for trade users.

A BUYER FOR EVERY SELLER

In the field of physics, most people are familiar with Newton's dictum "For every action, there is an equal and opposite reaction". (The recoil when a shotgun is fired is a frequently-quoted example of this). In the same way, it is essential for the LME's smooth operation that for every hedge seller there should be a buyer. Clearly, within the community of trade users of the Exchange there are likely always to be both hedge sellers and hedge buyers in the same ring (smelters and semi-fabricators, for example, who tend to react in opposite ways to an impending price trend). They can quite likely be the ultimate counterparty to each other. But this broad principle is subject to many detailed differences which can result in the market experiencing an excess of buyers over sellers, or vice versa, in any one ring.

According to the law of supply and demand, when this happens, there is no alternative but for the price to go down, if there is an excess of sellers, or up if there is an excess of buyers. It is at this point that it is important that scope in fact exists for non-trade users to come into the market and become the ultimate counterparty to the hedge buyers or hedge sellers who would otherwise have to remain unsatisfied. Needless to say, non-trade users of the Exchange are not a charitable institution, and their motivation for operating on the LME is not simply to help trade users do their hedging.

A non-trade user like a bank, for example, may well have financial requirements to satisfy which can most easily be met at the time by an operation on the LME. It should be food for thought for those in the trade who still insist that use of the LME is gambling that banks are Exchange users. For one attraction for the banks to perform their financial operation through LME trading is that this trade is, for their purposes, virtually riskless. Certainly it will be a lot less risky at that time than attempting to meet their financial objective another way. This will be explained in more detail later in the book.

The years 1994 and 1995 have, as already noted, been distinguished by the more extensive use of the Exchange (and other commodity futures markets) by investment funds. It is understandable that the motivation of such funds should be viewed by the purely trade user of the Exchange as speculative. If, therefore, he is already disposed to feel that use of the Exchange is speculation, he will no doubt smugly say that the presence of the funds on the market proves his point. Yet ironically, the presence of the funds on the market has been stimulated by their belief that the fundamentals of supply and demand in the relevant metals are not being correctly reflected by the prevailing price. In the sense that they are buying (or even sometimes selling) in anticipation of a correction in prices to reflect the fundamentals more accurately, they are speculating. Yet in the sense that they are trying to peer into the future and read the trend of the fundamentals, they are doing exactly the same thing as the trade!

Lastly, among Exchange users, we come to the individual speculator. A relatively rare animal these days, he may well in fact be operating on the Exchange only through money he has placed with one of the big investment funds. But those operators who continue to trade independently are performing a function quite similar to the investment funds. This is to say that they are trying to profit from anticipating a move, up or down, in prices in response to a change in the fundamentals that the market has not yet fully reflected. Additionally, because the private speculator may well be specialising in metals, perhaps capitalising on experience of the Exchange as a trade user, he will also be looking for short-term movements in price in response to technical factors. In this sense, he will be legitimately contributing to the market's liquidity, the importance of which will also be explained later in the book. Last but not least, the trade user of the Exchange disposed to be nervous of the presence of avowed speculators on the market, even if at a very low level of activity, should reflect on the following. Cynically, it is said that private speculators are always eventually net losers to the market. Now if this is true (and it is not the place of this book to comment on this), the speculator is making a contribution to defraying the trade user's broker's overheads!

CLIENT-BROKER RELATIONS

One of the important elements in successful use of the exchange by any client – whether or not a trade client, and whether for the purposes of hedging or otherwise – is his relationship with his broker. At one extreme, it can be argued that the client is giving orders, and the broker executing them. Therefore, the broker doing the job for the least money is the best one for that client. However, even for the individual speculator this view is not realistic. For the trade user of the market with a risk management programme to implement, it is downright dangerous.

It goes without saying that a trade client needs faultless and time-sensitive execution of his orders. On the LME, as distinct from some other futures markets, he also needs and expects some kind of credit line for his open positions. On a market like the LME where contracts are run to maturity, there is no immediate netting-off of positions, and making margin calls and margin payments strictly according to the book would require a flurry of cheques and administration that would be little fun for anyone. However, client credit to smooth cash flow on this account is a factor that has entered into some of the stories of futures trading programmes that have gone wrong in recent years. Those whose experience is limited to markets where netting-off and strict margin calls are the rule are apt to cite client credit as a cause of weakness. Conversely, those whose experience has been built up on the LME system find operations on other markets irksome. It is the hallmark of a good broker-client relationship on the LME that the broker shall have sufficient understanding of the client's objectives and modus operandi to "go with the flow" of the client's thinking. Indeed, in the vital service of keeping the client informed of macro and micro factors affecting the market, a good broker will know which elements are most important to the client's programme.

Above all, it is in the almost philosophical area of risk management strategy that there is the greatest need for the broker to be completely in tune with the client. Anyone reasonably expert in the use of the market can tell you (especially with hindsight!) about when it is a good idea to emphasise futures or options as the main risk management tool. However, when it comes down to the day-to-day choice of what to do next, it is easy to lose sight of the wood for the trees. At such a time, debate between the client and broker is most productive if it takes place between equals, both dedicated to achieving the client's objectives. Nobody will know better than the client the reality of his metal position, manufacturing programme and risk profile. But he cannot expect to know as much as his broker about technical developments on the market and last-minute developments. Nor can he expect to have as much experience as his broker of actually implementing different hedge techniques in different market situations.

Last but not least, not all of a broker's clients are trade hedgers. They will probably also include funds, banks and outright speculators. How can a broker be such a chameleon as to serve all these clients? Isn't the truth that he has one service programme, which he implements for all his clients? No, it's not like that. To start with, within a brokerage house there are specialists, and a client will be assigned a broker specialising in the kind of trading the client wants to do. However, the broker's breadth of experience of the reactions of all kinds of exchange user to all kinds of situations helps him advise the client of the marketplace realities against which the client – and his broker – are trying to implement the client's programme. The other constraint is one of the oldest in commerce. If the broker does not provide each of his clients with the service that client wants, he will fairly quickly run out of clients. Most big brokerage houses measure their big trade clients by tens rather than hundreds and the same applies to non-trade clients. So there is not much future in running a brokerage house on the basis that a lost client can be easily and speedily replaced.

The reality of the normal relationship between most clients and their brokers takes us back to the point made in Chapter 1; on a day-in, day-out basis hundreds of clients satisfactorily and constructively trade tens of thousands of tonnes of metal across the ring of the LME, the lion's share of it being trade hedging. Most clients would admit (except, of course, to their broker!) that a collaborative, expert and professional service from their broker is a key to their successful trading. And most brokers would admit that they get pleasure and stimulation from serving their client in a team relationship.

MENTAL APPROACH VITAL

Last but not least of the preparations necessary before embarking on consideration of hedging, or any other use of a futures market, is an acknowledgement that it is necessary to cultivate a suitable mental approach. Most people have no previous basis of experience for, as it were, living inside a world in which such a kaleidoscopic and continuous movement of prices is taking place. This is particularly true of many end-users of metal accustomed to long periods of stability in cost items like steel, oil or labour. True commodity markets like metals and soft commodities, as well as markets in equities and currencies, have been described as "designed to hunt". In other words, they resemble a dog following a scent. Over a distance, the dog follows the scent remarkably accurately, but to do this, it continually casts about either side of the central line to ascertain where the scent is strongest. With commodities, there is the added complication that nobody has previously laid a trail! The market is therefore continuously hunting about trying to find an equilibrium between all the various inputs from supply, demand and, often most importantly of all, "market sentiment". This expression connotes all the hopes and fears that are running through the minds of all those who go to make up the market.

Sometimes these sentimental factors are entirely understandable. For instance, a civil war is brewing in a major producing country. Will it occur, and if so when? If it does occur, will it affect production and exports as severely as was supposed when the issue first arose? Or will it be compensated by increased exports from a major producer or stockpile which has been holding off the market waiting for higher prices? How will the secondary sector of that particular metal industry react? From day to day the consensus view on these questions will certainly veer about. If the consensus on the war becoming a significant bull factor moves up from 40% probability to 45% probability, sentiment will push LME prices higher, without any discernible variation in the fundamentals of supply and demand in tonnage terms. Futures market prices are not only the most efficient mechanism for discovering the price levels that will, in economists' terms, clear the market. They are also often a valuable early warning system, which allows the more sluggish physical market to start responding to new circumstances in good time.

At other times, changes in sentiment can appear completely irrational, and quite often it turns out that they were. This is the market's equivalent of the tracker dog temporarily losing the scent. To a hedger, however, it is not so important whether the reasons for the market settling at a particular level on a particular day are good or bad. What is important is that he has covered himself against the risk of a price at that level harming his business.

A further element of mental preparation is necessary for those already familiar with the twice-daily stream of LME closing prices. This is to realise that the cash and three months quotations that emerge from the morning and afternoon trading sessions in the Ring are only the tip of the iceberg represented by the total envelope of LME prices. Every minute of every day, not just during the rings, the forces of supply, demand and sentiment are acting on each trading position from cash to 27 months. If a hedger is concerned about a risk position extending from, say, a month hence to 23 months hence, he cannot assume that the differential between these two positions, whether contango or backwardation, will remain constant as the key cash and three months positions move up and down. In appropriate circumstances, even the passage of an hour or so can see a move in one or both positions in which he is interested which could materially alter the size of the differential, or even turn it from backwardation to contango or vice versa.

Once again, such behaviour should not be viewed as the market being perverse. Rather, it should be viewed as an example of the flexibility and responsiveness of the market which can be so helpful to the hedger.

Hedging in theory

The gamut of situations which may call for hedge action is very considerable, and that is why two whole chapters of this book are devoted to it. This chapter outlines the principles that underlie the various hedgeable situations, while the one that follows gets down to the hard graft of following those principles through in practical examples. The reader is strongly advised to read these chapters first in the order in which they appear – later on, it will probably be found necessary to job back and forth between them as the practical examples are followed through.

This point is illustrated by drawing a parallel with learning a foreign language. One can appear to make great progress by concentrating on vocabulary, but real command of a language is only achieved by those who take the trouble also to learn the grammar. The principles of hedging set out in this chapter are the grammar of the language of futures. It is also worth pointing out that an error of grammar while talking or writing a foreign language probably costs you no more than a moment of embarrassment. An "error of grammar" when dealing on a futures market – probably under pressure – can cost a very great deal more than embarrassment.

Most people are familiar with the fairground adage "What you lose on the swings, you gain on the roundabouts". Conceptually, hedging with futures is a process of ensuring that what you may lose on the swings is exactly matched by what you gain on the roundabouts. But the corollary is that if, unexpectedly, you gain on the swings, you have now ensured you will lose the same amount on the roundabouts. If hedging with options is feasible and appropriate, then you take steps to avoid expected losses on the swings, while remaining free to benefit from the gains on the roundabouts (if any).

Various attempts have been made to give a succinct definition of hedging. As far as hedging with futures is concerned, the following has been found over time to take a lot of beating:

1. Hedging is the offsetting of a perceived price risk in one market by the deliberate assumption of an equal and opposite risk in another market.

Remembering that options are more and more being used in hedging programmes either alone or in conjunction with futures, the following definitions are also most important:

2. Hedging with options, where appropriate, gives protection against a market move in one direction, but does not insulate the hedger from a move in the opposite direction.

3. Hedging is the exchange of the known for the unknown.

For the purposes of this book, definition No 1 means that a perceived price risk in a physical metal market is offset by the deliberate assumption of an equal and opposite risk (ie an equal and opposite position) on the futures market. Economists speak of a hedge in which all the elements of the transaction are exactly equal and opposite as a "perfect" hedge. As we shall see in this chapter (and even more so in the next one), the "perfect" hedge is found much more easily in theory than practice. But it is an ideal that the hedger should always have in mind. For without a rigorous understanding of what constitutes a perfect hedge, it is not possible to determine how far the best practicable hedge is not quite perfect.

The variable elements that must be matched between the physical and futures position to achieve a perfect hedge are:

1. The commodity, including the exact form in which it is traded.
2. The price. This means that a perfect hedge is only possible when the physical commodity is priced on the basis of the futures market price (and, of course, in the same currency).
3. The time. This means that the futures market position must open and close on exactly the same dates that the risk position in the physical market is open.

It will already be clear why a perfect hedge is very rarely feasible, so let us pause to consider why thousands of hedge transactions go ahead every day when the hedger knows that his best practicable hedge is not quite perfect. To do this, we will draw a parallel with the insurance business. Almost all businessmen accept the need to insure their factory or office against fire risk. They may shop around in the market for the best-value cover, but will, at the end of the day, cheerfully pay the insurance premium without worrying too much whether they have achieved a "perfect hedge" by so doing. They do so because the effects of NOT having insurance, in the event of a fire, are so potentially disastrous for their business that they know it is a risk they must do their best to cover. Yet, as those who have had occasion to claim on insurance know, it very rarely provides "perfect" cover. The assessment of the value of the premises may be erroneous – it may even be outdated by a change in market values during the life of the policy. The effects on the business from a partial destruction of the premises may be as bad as total destruction, but the insurer will only pay out in proportion to the part of the premises destroyed.

So the hedger's thought processes are the same as those of the purchaser of insurance. Covering 95-99% of the risk of a completely unacceptable loss is infinitely preferable to exposure to 100% of that risk. Comparatively, the 1-5% of the risk that is not covered is commercially acceptable as part of the process of risk management, just as the insurance premium is. It is remarkable how people with only a sketchy knowledge of hedging will refer to a loss-making hedge position, or an abandoned option, as "a loss" even though it is, when correctly taken in conjunction with the related physical position, by design a zero-sum game or a positive gain. Yet these same people would never refer to the cost of a fire insurance premium as "a loss" just because the building never burned down! Even more unfortunately, in some countries, the tax man is among those whose knowledge of hedging is sufficiently imperfect that he may even try to claim profits tax on a profitable hedge transaction, while disallowing the loss on the physical position that the hedge has offset. So an important practical lesson from this theoretical point is that you need to understand not only hedging, but also your local tax collector's view of it.

Hedging as we have outlined it above appears to be an entirely reactive, not to say defensive operation. In situations where risk-avoidance is paramount, that is as it should be. But it is also possible to approach risk management in a more dynamic way than by simply trying always to offset it. We touched, earlier in this chapter, in definition No 2 on the place of options as a hedge tool which can allow the hedger to secure protection on the downside or upside (depending on the nature of his risk) without locking him out from the benefits of a move in the market in the opposite direction from that which he feared. This is certainly not purely defensive, but it is still largely reactive to an expectation of a market move. Let us therefore examine why we have given three definitions of hedging, and why No 3, is wider than, but not in conflict with, the first two.

Most businesses spend many hours each year, quarter or month working out budgets that will define how the business will operate in the next period. They will guess at what metal prices may do in the period and what this will mean to them. They will then, if they have taken this book seriously, put in place a hedge strategy designed to protect them as effectively as possible against the risks they see of prices moving to levels that will hurt them.

However, this is not the way they will budget the output of the factory. They will set targets in unit or tonnage terms that they desire to achieve, given the variables of capacity of machinery and men, supplies of raw materials and the constraints of a competitive marketplace. Only very rarely will companies budget also their desired metal price in the coming period, and look into ways of achieving this target. It may seem a relatively small distinction between deciding on a price trend that you would not like and nominating a price range that you would like. Yet this is what distinguishes reactive and defensive hedging from strategic and dynamic hedging. It is what distinguishes "exchanging the known for the unknown" from neutralising one risk with an equal and opposite risk.

It may be argued that budgeting a metal price for a future period is to risk comparison with the legendary King Canute, who ordered the tide to stop coming in – and failed. But if Canute had decided a position on the foreshore that he wanted to maintain, and had, in good time, built a small pier or jetty, he could have kept his position, kept his feet dry and saved himself the trouble of ordering the tide to do something he knew it wouldn't.

KNOWING YOUR VARIABLES

Although this chapter is dealing with hedging at the level of theory, there is one practical point that should be emphasised at this stage. This is that trade hedgers, because of their familiarity with their physical position, are more likely to make a mistake with what is to be hedged on the physical side of the equation than with the hedge transaction per se. An elementary example will illustrate this:

A miner has sold a boatload of concentrates cif and has hedged the value of the metal content for the period the concentrates are on the water. Due to storms or strikes or vessel breakdown, the boat is ten days late arriving. The miner routinely negotiates with the customer for late delivery of the material and deals with the related banking, insurance and stevedoring practicalities. He forgets, however, that his ideal is the achievement of a "perfect" hedge and neglects to roll his hedge forward by the requisite ten days. His existing hedge must then be closed out ten days before the arrival of the metal, but during those days the market falls steeply. This is exactly the risk the hedge was taken out to offset, but just when it was needed, it wasn't there.

It is no good blaming the hedge for this. It is a dumb tool, and cannot be expected to know about the shift in the arrival date unless the miner tells his broker and arranges to extend the hedge by the requisite ten days. Other things being equal (and the next chapter explains how sometimes they are not), the cost of this operation is minimal – certainly in relation to the size of the risk. Also, the LME's structure of daily settlements from cash out to three months always allows the timing of the hedge to tally exactly with that of the physical trade.

A key theoretical point for the aspiring hedger to grasp before he starts to grapple with the practical examples in the next chapter is that hedging is a process. Too much study of specific examples too soon can lead to the erroneous impression that hedging is a series of hammer-blows. The price starts to go up so – bang – the hedger knocks that on the head with a sell hedge if he is going to be short. Then it goes down so – bang – he takes out a call option to lock in the advantage of the prevailing trend just in case it goes into reverse while he is still short.

In fact, the skilled hedger's role model should rather be a snooker player or racing driver. Both of these know that successfully completing the next shot, or negotiating the next bend, is only part – sometimes the least part – of that action. What is critical is to complete that action in the most advantageous position to embark on the next shot or the next corner. In this way the successive actions flow together in a seamless stream and are the exact opposite of a series of hammer-blows.

Equally important is the need to understand that hedging is not a black-and-white situation, it is various shades of grey. We have already pointed out that the economist's "perfect hedge" is hardly ever possible in real life. Operating a hedge strategy costs money just like any other form of insurance. You have to pay for peace of mind. So the shade of grey that always has to be considered is the proportion of the value that it is desired to protect or achieve that will actually be protected after all costs have been met. It should be added that the accountants do not usually allow any great value to be put on the blood, sweat and tears of the hedge programme manager!

However, there is a piece of good news on this score. Sometimes, some hedge operations not only achieve the desired protection of a metal value, but as a by-product they produce a trading profit that will at least pay the brokerage on the hedge transactions, and may do a lot better than that. As will be explained in detail in the next chapter, such situations arise when one enters a sell hedge in a contango market which falls, or a buy hedge in a backwardation market which rises. These situations can be compared to an insurer paying the insured a premium for the privilege of insuring his risk! Needless to say, these situations do not arise every day, but when they do, they help to offset the costs of hedging in more normal or, worse, much more adverse circumstances. Also needless to say, such situations do not normally arise with hedge operations using options, since the option is either exercised or abandoned. However, there can be situations where a correct hedge strategy will involve selling options, rather than buying them and in these circumstances, again, the hedge process will itself produce a small revenue.

COMBINING YOUR INSTRUMENTS

In recent years a whole new approach to hedging has emerged which is usually referred to as "dynamic price risk management". It accepts the abovementioned point that hedging is measured in shades of grey and profits from the dynamic characteristics of options as Exchange-traded instruments. Conventional hedging solely with futures was designed either to offset a specific risk or to fix a specific price level. Risk management techniques, by contrast, are designed to protect their users against a defined range of detrimental price outcomes over a targeted period, while retaining the range of price outcomes which are advantageous to the hedger.

This type of protection is achieved by combinations of options and futures. For instance, a hedger who essentially wants to protect himself against a range of higher prices buys a call with a strike price at the bottom of this range. If he wants downside protection, he buys a put option with a strike at the top of the chosen range. In doing this, he of course incurs a premium cost; the challenge of price risk management is to minimise this cost. This is done, firstly, by buying his option when his chosen strike price is out of the money (above the present price for a call, below it for a put). Secondly he caps the range of call protection by granting (selling) an option at a higher strike call (or lower strike in the case of his put protection). He may even grant call and put options on a range of prices which is not important to him for the sake of earning the premiums. These combinations of options and futures are known by a variety of names, such as min-maxes, synthetic puts, participating calls, strangles, straddles and the like. The reader who is more interested in understanding than reciting jargon is best advised to break down the combinations into their components and analyse the specific protection each part brings.

This approach has the added advantage of being dynamic – able to reflect a changing set of management expectations and goals over time. The conventional futures hedger, once he has conceived and executed his hedge, is pretty much locked into his original decision; his strategy is a static one. An exchange-traded option by contrast is an insurance policy. True, it expires at a given moment of time, but, once bought, can be resold for its market value at any moment up to expiration. Over the course of time it frequently happens in business that the goalposts are moved by market circumstances or management preference. When this happens, the risk manager can change his level of cover, or the level of participation he has given away to pay for it; he can abandon the cover by selling the option to recover part of his premium costs; ultimately he can switch his insurance protection into futures when he is convinced he is safe to do so. In short, he can optimise his strategies over time.

Many hedgers first encounter risk management techniques under the guise of "swap" instruments. A swap is a device originating in the banking business to manage interest rate exposure. In a swap, a counterparty offers to the hedger to substitute for a given period forward a fixed price for a floating price.Or he may offer to substitute floating for fixed; in either case the swap is settled with a cash payment either way at the end of the period. Such arrangements, known traditionally as "averaging contracts" have been offered as a service by LME brokers for many years. They are used by hedgers whose metal intake or output is priced on LME average settlement prices and who wish to avoid the minutiae of the classic futures technique of hedging "at the close" every day in the averaging period.

Nowadays, alongside classic "European" options on futures, increasing use is made of Asian options. With these, the chosen strike price is liquidated automatically against the relevant month's LME average, if it is in the money, rather than by declaration. As this book goes to press it is expected that Asian options will be adopted as Exchange-traded instruments. Once again these instruments are best understood by the hedger breaking them down into their LME components and applying them to his specific business risk. Whether the hedge strategy is effected by a swap agreement or a combination of LME contracts becomes a matter of convenience against flexibility for the hedger. An off-Exchange swap agreement probably involves less administration, but an LME contract is more fungible – ie it can more easily be transferred to an alternative counterparty.

Now (and only now) it is time to turn from theory to practice.

CHAPTER 4

Hedging in practice

This chapter will attempt to cover all the normal combinations of risk management objective and market circumstance at the time of entering a hedge. It will also attempt to show differences of practical approach as between defensive and dynamic hedging. We say it will attempt to cover these variables, but in reality the permutations and combinations are almost limitless. There are bound to be situations that confront the client-broker team that do not conform exactly to the various scenarios that are covered here. This is when a firm grasp of the principles outlined in the previous chapter is essential to understand which of the examples given is nearest to what the situation demands. Having got a "best fit" hedging approach, it is then necessary to see if that is in fact good enough on a normal cost-benefit analysis, or whether a refinement of that hedge technique will get closer to the desired result.

One of the things a client should always beware of is falling in love with the technicalities of hedging. Most people know a manufacturing company which came to grief because the boss, or the Board, became obsessed with a new product or process, and pursued research into it to the point where they could never obtain a commercial pay-back from the initiative. It is also possible to become so obsessed with the techniques of hedging as to lose sight of the reason for doing it. There is only ever one reason for entering into a hedge, and that is to manage a risk situation. Completing a fiendishly complex hedge operation just to get into the "Guinness Book of Records", while losing sight of the risk situation to be managed is a recipe for disaster.

Most examples will also focus on the use of the first three months of the LME trading period, which is the period in which every business day is a trading day on the Exchange. However, there will also be a few examples showing how, in appropriate circumstances, use of the further forward dates (up to 27 months) is more cost-effective, though possibly slightly riskier because of the less liquidity in the more distant months than in the first three. Last but certainly not least, the use of options in preference to, or in conjunction with futures will be illustrated.

Example 1: FABRICATOR HEDGES LONG POSITION IN LEVEL MARKET

The classic example of a hedge, just to set out the principles involved, assumes some artificial circumstances. Chief of these is the assumption that LME prices are, throughout the hedge, level – ie, there is no backwardation or contango. In reality, one or other of these situations is present 99% of the time, and one or other tends to have dominance of a particular market for months or years at a time. A contango, let us recall, is where the spot or cash price is at a discount to the forward price. A backwardation is where the cash price is at a premium to the forward price.

We will take the example of an independent aluminium mill which has successfully tendered on a contract for the supply of 1,000 tonnes of aluminium roofing and siding for a factory project. He needs to buy the raw metal now to process over the next three months, but the contract requires him to supply the sheet on the basis of the aluminium price ruling at the time of its delivery. He is worried the price will fall against him in that interval. Against a tender necessarily geared to current metal prices, that would be disastrous.

Reverting to the concept of the "perfect" hedge set out in the previous chapter, let us examine the elements that need to be matched.

First, the commodity. As far as the metal that is at risk is concerned, the fit with LME futures is as near "perfect" as you can get. Maybe some material is supplied as ingot and some already in slab form, and therefore subject to a shape premium, but it is all unwrought, so values will move almost exactly in line with LME prices. True, as it is progressively converted to sheet, he must regard his metal at risk as subject not only to raw metal price changes, but also to the risk of changes in rolling margins. Unfortunately, there is no futures market in rolling margins. Fortunately, however, as he has tendered with only the metal price as an unknown, it is up to him to build into the quote a rolling margin he can live with.

Next we must look at the price. For an independent mill in Europe, this is a pretty open and shut case. His raw metal price will vary more or less exactly in line with LME movements and LME prices are quoted in the currency of international aluminium trading – the dollar. So again we have almost a "perfect" fit (except of course for the currency element in sales of product in the local currency – but this is a separate currency hedging issue, outside the scope of this example). In North America, Japan and some other markets, there is some lost motion in the linkage between LME prices and actual prices for unwrought metal in the local market – and this, again, is after ignoring for the moment any currency factor. This could indeed be a problem if the local market is controlled, such that he has to buy his raw metal locally, but normally, he can control the fit between his physical price basis and his hedging price by buying basis LME.

The third item to match up is time, and here again, in this example, his period at risk is three months, exactly equal to the period of maximum LME liquidity and of daily settlements. His "equal and opposite risk" hedge operation would then look like this:

Physical	LME price	Futures
Buys 1,000 tonnes at $1700 (cash price) ie he is long	Cash $1700 3m $1700	Sells 1,000 tonnes 3 months at $1700 – ie he is short
	3 months pass, Market falls	
Delivers sheet basis $1600	Cash $1600 3m $1600	3m sale at $1700 matures. Buys back 1,000 tonnes cash (spot) at $1600
	Result	
Tonnage +1,000 then -1,000 = balance. Value – LOSS of $100 per tonne		Tonnage – 1,000 then + 1,000 = balance. Value – PROFIT of $100 per tonne

As can be seen from this over-simplified example, a potential loss of $100,000 has been eliminated. In reality there is still the small cost of broker's commission on the purchase and sale transactions to be considered, but as an "insurance premium" to protect a risk of loss of $100,000, it is cheap.

Example 2: FABRICATOR HEDGES LONG POSITION IN CONTANGO MARKET

We will now strip away successive elements of the artificiality in the above example to get nearer and nearer to real life, whilst retaining, we hope, the conceptual understanding made possible by Example 1. In this example, we have eliminated perhaps the greatest artificiality of all – a market that remains level during three months. Classic futures market theory tells us that it is normal for futures markets to run at a moderate contango (cash at a discount to the forward position). This is because it is normal for the market structure to reflect the cost of carrying metal for three (or more) months, made up of the opportunity cost in interest on the money tied up, warehousing and insurance. As we will see later in the book, because of the standardisation of the warehousing process on futures markets like the LME, these last two items cost remarkably little. It follows that there is a limit on a contango about equal to these costs. If it were to go any wider, the world and his wife would buy cash and sell forward metal, locking in a profit greater than the actual cost of holding the metal for that period.

This is a good point at which to make clear that if a market is in backwardation (cash at a premium to the forward position), this does not mean it is being perverse or behaving abnormally. It merely means that supply-demand pressures on the spot position are producing a higher price than the corresponding supply-demand pressures on the forward position. In this context, the level of LME warehouse stocks and whether they are rising or falling is an important influence and indicator. A long-run study of postwar LME prices shows that in a trade-oriented market like the LME, it is not safe to assume that a contango is "normal" or that a backwardation is obviously a temporary aberration. The other important point to make about a backwardation is that, unlike a contango, it has no "natural" limit. Our contango market example looks like this:

Physical	**LME price**	**Futures**
Buys 1,000 tonnes spot at $1700	Cash $1700 3m $1750	Sells 1,000 tonnes 3m at $1750
	3 months pass Market falls	
Sells 1,000 tonnes of sheet basis $1600	Cash $1600 3m $1650	3m sale at $1750 matures. Buys back 1,000 tonnes spot (cash) at $1600

(example continued)

	Result	
Value – LOSS of $100 per tonne		Value – PROFIT of $150 per tonne

This shows that exactly the same operation as in Example 1, when carried out in a contango market, produces a profit for the hedger of $50 per tonne, or $50,000 on his three months' period of exposure. This very handsomely outstrips the brokerage cost of the futures operations.

Example 3: FABRICATOR HEDGES LONG POSITION IN BACKWARDATION MARKET

In the interests of balance, we will now re-run this example once more, but this time in a backwardation market. In this example, the artificiality is that the backwardation is shown as steady throughout the three months. This can happen, but the situation to be most feared by all short hedgers (as in both our examples so far) is a widening backwardation. The risks of a moderate backwardation widening are not always easy to assess, and the speed at which it is capable of opening out to truly alarming proportions can sometimes catch even the most experienced hedger on the wrong foot. However, it is no good saying "Never short hedge in a backwardation in case it suddenly widens out ." The hedge you did not enter for fear of a widening backwardation could turn out to be the one hedge in your life you could not afford to omit! Rather, that is the time for considering alternative hedge strategies to the plain short hedge considered thus far.

Physical	**LME price**	**Futures**
Buys spot at $1700	Cash $1700 3m $1650 3 months pass, Market falls	Sells 3m at $1650
Sells sheet basis $1600	Cash $1600 3m $1550	3m sale at $1650 matures as spot. Buys back spot at $1600
	Result	
LOSS $100 a tonne		PROFIT $50 per tonne

The overall loss of $50 a tonne is exactly equal to the backwardation at the time of opening the hedge – that is the known element, whereas the unknown is the extent of the drop in market prices. This is likewise why the futures transaction profit in the contango example was $150. After the $100 movements on each side of the equation are netted off, the $50 on-balance profit was assured at the time of entering the hedge because that was the size of the contango. In the context of exchanging the known for the unknown, paying a backwardation of $50 a tonne can still look smart if the market falls $100 or $200 over the period. And it happens.

Example 4: FABRICATOR HEDGES LONG POSITION IN BACKWARDATION MARKET II

We will stay with our mythical independent aluminium sheet mill once more to illustrate a possible alternative response to the scenario of falling market and backwardation. Because of the uncertainties associated with a backwardation – the possibility of a "squeeze" on the nearby position sharply widening the "back" during the life of a hedge – the mill could turn to options as the means of securing its protection. As a short hedger (one who is hedging a long physical position), he buys a put option, giving him the right to sell, as this has the effect of making him short up to the moment of exercise of the option. Still using the same price scenario, it is fairly obvious that if the mill fears a fall in prices, so do a lot of other people. Option premiums, it will be remembered, are subject to their own supply-demand pressures. Although seeking to buy a put option, with the current price of $1650 as the strike price is likely to be expensive, because it is "at-the-money" (after all prices may not fall the way he fears), it may still be within a controlled cost budget. However, it may be possible to choose a strike price "out-of-the-money" at a more acceptable premium. This example chooses an "at-the-money" strike and would work out as follows:

Physical	**LME price**	**Option**
Buys 1,000 tonnes spot at $1700	Cash $1700 3m $1650	Buys 3m put on 1,000 tonnes at strike price of $ 1650 for say $25 per tonne
	3 months pass, Market falls	
Sells 1,000 tonnes of sheet basis $1600	Cash $1600 3m $1550	Exercises option to sell at $1650 and buys back spot at $1600
	Result	
LOSS of $100 per tonne		PROFIT of $50 per tonne LESS the option premium of $25 = $25

The net loss of $75 per tonne is actually a worse outturn than in the previous example. Also, if the market had risen, participation in additional profits would have been available by using options. But, if the market had not fallen at all, contrary to expectation, the futures hedge would still have shown an on-balance loss of $50, whereas the option strategy would have resulted in the option being abandoned and the on-balance loss being its cost, namely $25 a tonne. This illustrates how options are more flexible than futures, but are certainly not a cure-all.

Example 5: END USER HEDGES SHORT POSITION IN CONTANGO MARKET

We will now look at the opposite side of the hedging coin – a constitutionally short position in physical which is protected against a feared rise in price by a long hedge. The most obvious potential for such a situation, as we outlined in Chapter 2, is an end user. An automobile manufacturer, for example, is not able to vary the selling price of his finished vehicles to respond to increases in price of his raw materials such as aluminium castings. He is therefore rightly fearful of any uptrend in aluminium alloy prices, and protects himself against this by a long hedge.

At this point we should also strip away a further element of artificiality in the examples to date, which have shown one-shot hedging actions covering a three month period. As was explained in Chapter 3, hedging is not like that; it is a continuous process. A good hedger's evolving metal position and price exposure should be protected like the back of an armadillo with a series of overlapping hedges, the sum effect of which is his total futures position. Each of these positions, as we shall see as this chapter evolves, is capable of being lengthened or shortened and thus subtly altering the hedger's total hedged position. Thus there is no inconsistency in a secondary aluminium hedge programme which is generally framed in terms of three month contracts (where the greatest liquidity in LME trading is to be found), even though the alloy ingot-castings-end user marketplace mostly contracts for physical material on the basis of monthly deliveries and even though the automobile manufacturer has probably got to fix selling prices for a model year.

It will be realised that most elements of a long hedge programme are the inverse of the same elements in a short hedge programme. Thus a short hedger fears a fall in price (because he is long of physical), while a long hedger fears a rise. Equally, it is to be expected that whereas a contango favours a short hedger, a backwardation favours a long hedger. When a motor manufacturer makes his commercial plans for a model year, he forecasts, among many other things, his purchase price for aluminium alloy castings throughout the period. Nowadays, movements in the LME aluminium alloy price will be the main determinant of movements in his invoice price for castings, and a sharp rise in these prices over the model year will be distinctly uncomfortable. So he locks in a known figure by hedging his year's purchases. This will be done by a series of hedge transactions profiled to match the varying rates of tonnage deliveries of castings he has planned with his suppliers. One such transaction would look like this:

Physical	**LME price**	**Futures**
Orders 5,000 tonnes of castings for delivery in 3 months, basis price ruling	Cash \$1600 3m \$1650	Buys 5000 tonnes 3m at \$1650
	3 months pass Market rises	
Receives castings basis \$1700	Cash \$1700 3m \$1750	3m purchase at \$1650 matures Sells as spot at \$1700
	Result	
"LOSS" of \$100 per tonne		PROFIT of \$50 per tonne

With the hedge his loss is fixed at the size of the contango. Without it, he would be at the mercy of a rise in the market, however great. Furthermore, unlike the backwardation feared by a short hedger, the contango cannot get out of hand.

Example 6: END USER HEDGES SHORT POSITION IN BACKWARDATION

Re-running the above example again in backwardation gives a balanced view of the scope for this end-user:

Physical	LME price	Futures
Orders 5,000 tonnes of castings for delivery in 3 months, basis price ruling	Cash $1650 3m $1600	Buys 5,000 tonnes 3m at $1600
	3 months pass, Market rises	
Receives castings basis $1750	Cash $1750 3m $1700	3m purchase at $1600 matures. Sells as spot at $1750
	Result	
"LOSS" of $100 per tonne		PROFIT of $150 per tonne

Once again, the on-balance profit of $50 per tonne handsomely covers brokerage costs. And although he has gone long of futures, he does not even have to pay the warehouse charges on the relevant warrants as he never physically takes possession of the metal. As we saw in the case of the short hedger in a contango (Example 2), when a futures hedge not only provides the desired protection, but also produces a profit, it is unlikely that a strategy using options as a straight alternative will produce a better result. This is particularly the case with a long hedger in a backwardation, as the probability of a more volatile cash price in a backwardation will make an option more expensive to buy.

Example 7: MINER EXTENDS SELLING HEDGE CLOSE DATE BY BORROWING

We will now consider the (we hope) imaginary miner we cited in Chapter 3 who omitted to adjust the length of his hedge to keep in balance with a change in the close date of his physical transaction. Here's what he should have done.

As a short hedger, the miner extends his existing position by what is known as borrowing. This is short for "borrowing from the market". It means simply that he closes his short hedge by buying back his short forward position as spot, but at the same time opens another short position for the requisite period (in this case ten days). This he closes, just as he did the original hedge, by once again buying back as spot his matured forward sale. Obviously a transaction of this length is not as efficient in itself as one extending over some months, as the same amount of brokerage may be paid on the round turn borrowing transaction as on the original hedge. However, the penalty for being uncovered for ten days, in the event of the market falling suddenly, is just as great as for being uncovered for three months of insidious decline in prices. We pick up the thread at the moment of the original hedge date reaching maturity:

Physical	**LME price**	**Futures**
Concentrates fail to arrive on due date	Cash $2600 10 days over cash $2610	Original 3m sale matures; buys back equal tonnage of cash. Simultaneously sells same tonnage again for 10 days over cash
	10 days pass Market falls	
Concentrates delivered cif	Cash $2550	Sells 10 days over cash position for $2610 and buys back spot at $2550
	Result	
Original hedged position preserved through subsequent market fall of $50		PROFIT $60

Once again, because there is a contango over the ten days for which the borrowing action was required, an on-balance profit of $10 is realised after cancelling out the $50 market fall and corresponding profit within the futures transaction. This contango applies solely to the ten days with which we are concerned here. It is possible (though not very likely) that the next ten days (eleven days over cash to twenty days over cash inclusive) would then be in a backwardation. Certainly, it would not be unusual to find a cash – 10 days over cash contango within an overall backwardation between cash and three months.

It is also important to make clear that this operation deals solely with the ten days for which the miner would otherwise be uncovered. It has no impact on the original hedge. If that was a short hedge that was also entered in a contango market, the miner may well feel pleased with himself on all counts. But if he had entered a short hedge in a backwardation, and the market had unexpectedly risen instead of falling, he might feel sorely tempted to assume that the market could not fall, in the ten days his physical delivery was delayed, to below his original hedged price. In that case, he might also be tempted to save the brokerage on a further hedging operation. The agony of that decision would have to rest entirely with him.

Once again, there is a symmetrical opposite risk situation and corresponding hedge approach. If a long hedger needs to extend his position, he will lend – ie lend to the market. And as part of the same symmetry, this operation will be favoured by a backwardation. All this can be distilled into two simple rules:

To extend a sell hedge (short futures), you borrow, and such operation is favoured by a contango.

To extend a buy hedge (long futures), you lend, and such operation is favoured by a backwardation.

Example 8: FABRICATOR EXTENDS BUYING HEDGE BY LENDING

Let's check it out by re-running Example 7, assuming a fabricator who has booked a delivery of raw metal for three months ahead basis price ruling and who wanted to protect himself against a rise in price meantime with a long hedge. We also assume a nearby backwardation is running. As delivery day approaches, his supplier invokes force majeure (diversion of vessel because of a dock strike) and assures delivery one week late. Our fabricator, still fearful of a price rise in that week, acts as follows:

Physical	**LME price**	**Futures**
Force majeure declared	Cash \$2600 5 days over cash \$2590	Sells original 3m purchase as cash at maturity. Simultaneously buys same tonnage of 5 days over cash
	5 days pass Market rises	
Metal delivered	Cash \$2650	Sells 5 days over cash position at maturity
	Result	
Original hedged position preserved through subsequent market rise of \$50		PROFIT \$60

In this case, of course, $10 of the "profit" on the futures trade over five days is due to the backwardation in force when the lending operation was initiated, with the other $50 going to offset the "loss" due the rise in price of $50.

Another circumstance in which carrying to extend or bring forward the close date of a hedge transaction is routine is where a long-term hedge has been entered into, using the LME's ability to trade up to 27 months forward. It will be remembered that trading further forward than 3 months is done on the basis of weekly and then monthly, rather than daily settlements. In entering, say, a 27 month hedge, the hedger may know at the outset his exact desired close date, and it does not correspond to the settlement date in the relevant month. More likely, he will not even know the exact close date his physical risk position will produce until he gets nearer to that date. Either way, he must expect to adjust his original 27 month hedge by borrowing or lending as it nears its close. Any thought that this situation is onerous should be tempered by recalling that, previously, protection over this length of time could only be secured by rolling one three month hedge over into another three month hedge. The brokerage costs for rolling over eight times are not nice to contemplate – at least for the client. And this is without considering the many more dates on which the vagaries of the market would be likely to throw up margin calls under the old regime.

A further important point to be made, now that we are talking about fine-tuning hedge periods down to the last day, is that an LME "cash" (spot) transaction is in fact delivered two working days later. This is because the LME is a trade-oriented market, and all transactions are capable of being settled with physical delivery. At all times, therefore, the cash transaction must be made two days prior to the prompt date on the forward position that is being closed out.

Only in this way will a cash purchase, for example, correctly balance a maturing forward sale.

At the same time, the growing use and increasing sophistication of options as a hedge mechanism means that these are also being used in preference to futures in many cases where a long-term hedge is desired. Apart from their greater flexibility, as mentioned earlier, options have the merit of a known up-front cost which eliminates the need for margin calls. Conversely, the further forward an option extends, the greater the uncertainty the grantor faces and the bigger the premium.

Another situation where either futures or options could be the best hedge medium is one frequently encountered in physical metal trading – the monthly average price-taking period contract. Under such contracts, deliveries of metal in a month are priced on the basis of the average LME price of that metal in the month of delivery – or quite often in the month before or month after the month of delivery. In this case, the type of option favoured may well be what is called an Asian option which is settled on the basis of the average price for the month – or in some cases even a quarter or a year – for which the option was taken out. As always with options, the technique is simple, flexible and possibly less strain on the cashflow or use of credit, because of the absence of margin calls. But there are still many situations where the traditional way of hedging an average price exposure with futures is still hard to beat.

The traditional technique, in simple terms, is based on assuming that the month's tonnage is delivered in equal instalments each working day. This is not what happens on the physical market, but it is the mirror image of how the reference price for that tonnage is built up as the month's average price evolves. It is therefore the best way of simulating "perfection" as far as the price component of the hedge is concerned. And of course the total tonnage is the same on both the physical and futures sides of the equation at the end of the month. There are usually about 22 trading days in a month, so an intake of less than 550 tonnes of aluminium or copper, or 132 tonnes of nickel, cannot be neatly divided into a lot a day; in this case, and in any case where the daily unit is not exactly a warrant, a little ingenuity is needed to get as close as possible to emulating the process of building up the price average. Most brokers provide an averaging service to cover these smaller tonnages. In the following example, the hedger is lucky enough to operate with almost perfect accuracy by hedging two warrants a day.

He should not assume that once the month's programme has been set up he can leave it to his broker to tick off each day's transaction, as there may be unexpected developments in the month that call for an immediate response. His supplier may declare a force majeure shortfall on delivery, and as this is a force majeure the tonnage cannot be considered as "owing" at the correct monthly average – it is tonnage lost. So the hedge programme must be adjusted to keep the tonnage hedged equal to the tonnage supplied. If the market is, in practice, fairly flat that month, it may be safe simply to suspend the hedge activity as soon as the tonnage actually delivered has been matched. But if the market is moving rapidly in the month, simply stopping the hedge programme half way through the month will not match the average price on the invoice for the tonnage that has been supplied, as this will be calculated, per the contract, on the average of every day in the month. A borrowing or lending transaction equal to the tonnage already hedged would be one way of adjusting its value to the full month's average.

In our following example, we will assume normalcy.

Example 9: CUSTOM SMELTER HEDGES WORK IN PROGRESS EXPOSURE ON CONCENTRATES RECEIVED UNDER PERIOD PRICING CONTRACT

For this example, we think of a custom smelter of copper with a regular supply of concentrates under a period pricing contract. Just as his material inflow is, in principle, continuous, so is his hedging programme. As was outlined earlier, it is the net imbalance on his metal position that represents his exposure, but for simplicity's sake, we will assume that for the month in question his exposure is exactly equal to the tonnage of concentrates he is receiving, none of which can be delivered as finished metal for at least a month from receipt. We will also assume that the tonnage involved is 1,100 tonnes of contained copper and that there are 22 trading days in the month. He is, of course, long of this work in progress tonnage, so protects it by a sell hedge against a fall in the market before he can sell the corresponding refined copper.

Each trading day in the month, he will sell two lots of three months copper. Remember that his whole month's hedge programme is itself part of a continuous hedge programme with a series of overlapping 3 months contracts, any one of which can be lifted, closed early or extended to keep balance between his total exposed position and his total hedge. Continuing to assume that his tonnage exposure in the month is equal to his concentrates receipts in the month, he will build up a total hedge position in the month of 1,100 tonnes. He now knows that whatever happens to LME prices between receipt of the material and sale of the resultant metal, its value to him will be pegged at the average of the LME price in the month of receipt. This is certainly exchanging the known for the unknown; it could almost be classed as exchanging the known for the unknowable! This arrangement also sits well with the likelihood that a major part of his resultant refined metal sales will be to long term customers and also priced basis the monthly average.

Example 10: CUSTOM SMELTER HEDGES WORK IN PROGRESS EXPOSURE II

That is the traditional approach. Now let us look at the use of options to achieve the same cover. As the exposure situation we are looking at here is based on an intake of material priced on the basis of the LME monthly average, it is obvious that, operationally, an Asian option is a neat way of achieving the desired cover. Two words of warning are necessary in this connection. First, Asian options are strictly cash-settled, like a financial futures contract. There is therefore no possibility of delivering or accepting metal to settle an Asian option. Connected to this is the fact that, at the time of preparing this book, Asian options had still not been adopted as official LME contracts, and can therefore only be offered by brokers as principal-to-principal transactions, which are not cleared through the Exchange's clearing system. Indications are that they will be adopted in the not too distant future

The use of an Asian option is conceptually very simple. To hedge the custom smelter's long position he buys a put. In a rising market, he would achieve effectively the same cover by selling a call, but for the purposes of this example we will adhere to the very broad general advice that trade users of the LME are buyers of options for hedging purposes, and it is professionals – brokers or bankers – who write them. We then need to remember that the longer an option extends, the more it costs. So the custom smelter would probably buy an at-the-money Asian option put for the end of the month following the month of delivery of the concentrates. This second month is the period in which he is most likely to sell most of the metal resulting from this particular intake of concentrates. If the market falls during this period, his option will be well in-the-money against the daily average of the life of the option, and it will be automatically declared. At this point (or to be exact a couple of trading days later) he will be paid out the difference. If, on the other hand, the market rises during the period, the average for the second month, at which he is likely to sell the resultant metal, will be greater than that for the first month (on which he prices the concentrates intake), his physical position is profitable and his option is automatically abandoned.

Example 11: CUSTOM SMELTER HEDGES WORK IN PROGRESS EXPOSURE III

Last but not least, we should consider the use of European options in the same scenario. European options, unlike Asian options, are Exchange-traded instruments and cleared through, and therefore backed by, the clearing. Each trading day it is simple to calculate whether the option is then in-the-money or out-of the-money – one simply compares the strike price with the day's price. This makes the European option itself eminently tradeable throughout its life. On the other hand, European options are exercised once a month on the first Wednesday of the month for value date on the third Wednesday. This can often make the third Wednesday itself subject to abnormal pressures, resulting in an abnormal price. This is especially the case if an attempt at manipulation of the underlying market – for instance a squeeze – focusses on a third Wednesday. Even so, things are not as complicated on the LME as on several US exchanges, which trade on the basis of monthly rather than daily settlements. Here there is a convergence on the third Friday of the last month of each quarter of settlement of futures, options and options on futures. This is known as the "triple witching hour", which can sometimes be very bewitching indeed! A further valuable feature of LME European options is that they are regularly traded in US Dollars, Deutsche Marks and Japanese Yen, thus avoiding the need for a separate currency hedge.

Once again, our custom smelter achieves his long hedge by buying a put for the same tonnage as his exposure – ie his month's concentrates receipts; and once again we will assume that he buys it at-the-money on the first trading day of the month of receipt of the concentrates. In the case of a European option, his declaration day will be the first Wednesday of the second month following the month of the receipt of concentrates. He will probably reasonably expect to have sold all the resultant metal by about that time. If the market falls during these two months, he will exercise his put and achieve a satisfactory hedge, except for the cashflow implication of waiting until the third Wednesday for settlement and to find his exact settlement price. However, he can always sell his option as an in-the-money asset at any time up to declaration day, and if fine tuning is still needed to cover residual tonnage this can be done by borrowing.

Example 12: PRODUCER MANAGES REVENUE RISK DYNAMICALLY

Let us imagine a copper mining company in 1995 which looks at a current forward copper price of $2500 per tonne. This level is well inside the top quartile of the copper price cycle of the last fifteen years, but at the same time is well below the highs which have prevailed for significant periods of time. The company enters its two year planning period forecasting the likelihood of rapidly declining prices, with an outside possibility of a rapid economic recovery proving it wrong. It wants to cover itself against the risk of lower prices impairing its ability to service a debt. At the same time it does not want to be excluded from a bonanza period of high prices. Although this is rare, it is the dream of all mining investors. Above all, two years ahead of production it does not want to be locked into a commitment which it cannot easily adjust if unexpected developments occur. To achieve this with options it will have to pay a premium, but it has the opportunity to manage this premium cost down over the two year period.

The company buys a strip (a time series) of Asian put options with a strike price of $2200 and in quantity equal to the expected production. These options expire in each of the calendar months of the budget plan for 1997, two years forward. Because this strike price is so far out of the money, the premium cost is low. The company has now preserved its opportunity to participate in any price rise which may occur two years hence. The Asian puts will be liquidated against the average LME settlement price for the expiration month, the price the copper smelter will pay for the copper content of the concentrates. The broker will pay the miner the difference if the settlement average at the time is lower, but if it is higher the option is automatically abandoned. The miner then enjoys the higher smelter payments less only the small option premium he has paid.

Let us imagine that in one year's time the forward price has risen by $500 to $3000 per tonne. The company can now choose to roll up its protection level from $2200 to $2700 by selling its original strip of puts and simultaneously buying a strip at the higher strike price. Because the new strike is still well out of the money the premium cost of this trade-up is minimal. The mine can even pay for the cost of this improvement in one year forward cover out of its windfall profits from the run-up in prices. With the backwardation which commonly occurs in times of shortage, it may well be enjoying an LME cash price of $3500 for its current production. It would roll up its cover by granting nearby call options at $3600 and $3700. If these are called away, the prices are sufficiently attractive that the cap on revenue will be acceptable. Also the expiration date is close enough that the likelihood of a monthly average price seriously exceeding the call strike is low. In either case the credit of the premium income is an attractive reduction in the cost of the long term insurance.

Example 13: END CONSUMER MANAGES COST RISK DYNAMICALLY

We will now imagine we are an aluminium wheel manufacturer faced with the risk of designing and marketing a new wheel with an expected two year life cycle. As a high duty component, it will be made from a special, quite likely virgin-based, alloy; so hedging will be done in aluminium, not aluminium alloy. Our market analysis tells us that a change of 10% up or down in aluminium costs is acceptable to our margins; that a cost increase of between 10-25% would have to be absorbed for six months but could then be passed on to our customers; and that a cost decrease of this order could be enjoyed for twelve months before the customer would insist on an adjustment to wheel prices. The price sensitivity of the second year of the product cycle would not be apparent until negotiations with the customers were completed at the end of the first year.

If forward aluminium prices were $1600, the wheel manufacturer would buy a strip of calls for the subsequent twelve calendar months with a strike price of $1760 and grant a strip of calls at a $2000 strike for the second six months of this period. He would grant a strip of $1440 puts for the entire twelve month period. This hedge position would exactly match the market profile of the risk he had analysed. If the timing of his customer response to market change proved to be faster or slower than he had estimated he could simultaneously roll forward the expiration dates of his cover to a later date or back to an earlier date. He would do this by simultaneously selling the existing calls and buying the same strike for a further forward (or more nearby) expiration month. Likewise, if the customers proved to be more or less price-sensitive to changes in the market than anticipated, he could adjust the cover levels by rolling up or down to higher or lower strikes. When the new negotiations came up, he would start the process for the second twelve months all over again.

Both the above examples illustrate the point in Chapter 3 about the difference between nominating a price you would like, rather than protecting yourself against a trend you would not like.

CONCLUSION

Although our specific examples of approaches to hedging have gone into double figures, this is still no more than an outline of the approaches that can be taken. To start with, in line with what was said at the beginning of this chapter, there are many permutations and combinations of the elements set out here. Almost everybody with exposure to physical metal risk can envisage himself in the position postulated for each technique. Miners may only rarely be short and end-users only rarely long, but in between traders, custom smelters and fabricators can easily find themselves long at times and short at others, and should canvass all the techniques relevant to that situation. Then, in addition to the examples given, there are yet further "mix 'n match" possibilitites involving both options and futures.

Last but not least, there can be times when the market is in such a position that the best hedge that can be applied is still far from perfect. At such time the potential hedger is quite entitled to call upon his reading of the market and decide that no hedge action has a better risk-reward profile than any hedge action he can think of. If his broker is responsible and reads the market the same way, he will agree. If the market then does something really way out, and a positional loss results, this can still have a positive value. It can reassure the client that the effort he has put into hedging himself the other 99.5% of the time has been well spent!

CHAPTER 5

The other side of the market

In this chapter, we consider the use of the Exchange by all those who are not hedging a physical metal risk position. As was briefly explained in Chapter 3, non-hedge clients of the market are a vital ingredient in assuring the balance and liquidity of client orders which are necessary to assure the clean execution of all orders, whether from hedgers or non-hedgers. A futures market whose only clients were trade hedgers, however trade-oriented in its structure, would from time to time run into real problems of liquidity and balance at any time that the longs and shorts among trade hedgers did not more or less exactly balance each other.

It is, therefore, inappropriate for trade users of the market to sneer at, or complain about, non-trade clients. However, this did not prevent a fairly loud and sustained outcry in 1994, when international fund managers poured millions of dollars into the market in support of their belief that metal prices were fundamentally and cyclically undervalued. The fact that some of these funds are known as hedge funds only rubbed salt in the wounds of traditional trade hedgers, confused by the short-term gyrations that inevitably accompanied the actions of such large players. (Hedge funds began to be so called in the seventies and early eighties because, like managed futures funds, but unlike equities-based funds, they regularly sold short as well as buying long. However, their motive is speculative gain, not risk avoidance). It was also galling to trade operators on the market to see that the resources available to these funds were so huge that they could, and sometimes did, initiate a self-fulfilling prophecy with their massive operations.

However, on a year-in, year-out basis we should first consider some of the other professional users of the market such as banks, pension funds, managed futures funds and what are known as HNWI's (high net worth individuals). These last are likely to be the only true speculators in this group, as the managed futures funds, like the hedge funds with which they are sometimes (understandably) confused, should properly be described as system traders rather than simple speculators, albeit leveraged ones.

BANK ACTIVITY

We will start with the banks, with whom most trade clients of the LME and brokers have, of necessity, pretty friendly relations. Despite their powerful move into what is loosely called "derivatives" in recent years, banks still undertake a considerable volume of true banking operations on the LME. The most obvious of these has already been touched on – namely profiting from a contango which is greater than the actual cost of carrying metal for that particular period. Such a contango is said to be at "more than finance levels", and will not necessarily occur neatly between cash and three months. Banks (and particularly brokers on behalf of banks) regularly track the size of contangos on all kinds of futures markets. But few equal the scope on the LME for the chance of spotting a contango at more than finance levels between any two dates in the next three months, or between any date in the next three months and any other settlement date out to 27 months. With the size of funds at their disposal, banks soon mop up any opportunity of this sort, and one transaction, perhaps involving millions of dollars, will suffice to pull the two dates back into line and eliminate the opportunity.

Like trade hedgers, banks have increasingly appreciated the alternative route to achieving an objective in the market that is provided by options. On the LME, as already mentioned, an additional attraction of the Exchange-traded European-style options that are available is that they can be offered in four currencies – US$, Sterling, DM and Yen. On the other hand, due to the extremely narrow margins that normally prevail in riskless operations to capitalise on a no doubt ephemeral contango at more than finance levels, you need to know exactly what you are doing when using an instrument that is settled only on the third Wednesday of each month, and is not tradable for two weeks before that. The matter of Wednesday settlements is, of course, not a problem for futures transactions profiting from a contango at more than finance levels in which the forward position is further forward than three months. In this case, the relevant Wednesday eventually matures as the prompt date, even though the month has moved into daily settlement once it is nearer than four months forward.

The use of complex mathematical formulae to calculate risk on options

– such as the famous Black-Scholes formula – has encouraged the incursion into futures and derivatives markets of what are colloquially known as "rocket scientists", rare individuals with the ability to deploy and develop these formulae to meet trading objectives. If employed by funds, they may well be looking for an original approach to a speculative operation. But if employed by banks, they are more likely looking for ways to use futures, or options on futures, to hedge an innovative over-the-counter (OTC) option product they have tailored for a particular client. In this sector of their use of the LME, the banks are undoubtedly engaging in derivatives trading and accepting a degree of risk. As highly sophisticated professionals, they are as likely to be sellers of options as buyers.

High net worth individuals are something of a mythical beast. Everybody is after them – equity brokers, futures brokers, pensions and life insurance salesmen, bond dealers, real estate dealers, investment capitalists, fine art dealers, various sorts of sellers of Brooklyn Bridge and gold-diggers. In reality, they are few, mostly ultra-conservative and backed by the top-notch professional advice they can certainly afford. Nevertheless they are found acting on the London Metal Exchange, though few of them with the impact of Nelson Bunker Hunt, when he was attempting to corner the world silver market in 1979. (The LME had a silver futures contract in those days).

All of them are basically speculative in their objectives and are therefore assuming the risks that hedgers want to offset. However, their approach to dealing may range from a simple punt on a three months (or further forward) position on a metal about which they feel bullish, to quite sophisticated, low-risk/modest return technical operations, quite possibly involving options as the sole or partial vehicle for the transaction. Either way, they are forming part of the rich tapestry of orders that flow into brokers' offices each day, the execution of which is what drives the market and discovers the prices.

THE ROLE OF FUNDS

Managed futures funds have been active in all futures markets in tangible commodities for longer than the hedge funds. Indeed, their origins in the USA some twenty years ago predate the introduction of most of the financial futures markets which now form a large part of the canvas on which managed futures funds operate. Today, such funds are based not only in the USA but also in the UK and some other European centres as well as in Japan and the Far East. These funds are normally traded by individual commodity trading advisors, or CTA's, and their secular performance over the past ten or twenty years the more senior funds have been in existence has been impressive. Because of their roots in the USA, and because of the popular tendency to think of gold first when metal futures are mentioned, managed futures funds are probably more likely to be found operating on Comex than on the LME. However, this is changing, not least in the wake of the strong movement into base metals by the hedge funds in 1994. For their purposes, the LME's greater closeness to the physical market has made it their preferred arena. There is now debate as to whether managed futures funds and hedge funds are converging, perhaps to the point where they will become one class of fund.

As things are at present, hedge funds are definitely in a class by themselves, if only because they are so relatively few and generally powerful. Their ability to analyse markets for fundamentals for long-term positional trading is demonstrably formidable, and they have also shown their ability to adapt quickly to the mix of technical factors and sentiment that frequently drives short-term movements on the LME. It is said that "the onlooker sees most of the game" , and the very fact that hedge funds are looking into the metal markets from the outside may occasionally give them a better view of long term trends than most people in the trade who are living with the market day by day. An attractive feature of hedge funds' interest in trading base metals is that their size means that there would be little popular backlash against the markets if they should lose money on them. This is the reverse of the situation if a small man takes a punt on the market and loses it. The popular view is always that the big boys of the Exchange must in some way have cheated him.

When LME trading volumes in a year can represent many times total world production of that commodity, it is vitally necessary that the market should have the liquidity to handle this trading demand without being distorted by it. In this respect, the intervention of the funds on the market can only be welcomed. Nor should the scale of this intervention be misrepresented. Whilst, on some days, the funds are undoubtedly the biggest force in the market, it remains true that, up hill and down dale, they usually account for only 10%-15% of total activity.

Clearly it is advantageous to non-trade operators to use a cleared Exchange with high liquidity and to trade in "paper metal" in regulated warehouses which is instantly saleable. They also benefit from leveraging the bulk of their transactions by trading on margin, though they are often puzzled by the LME's unique requirement to run a position to maturity before settling the contract, or netting it off against other parts of the client's position. All this makes it clear that non-trade users are "doing their own thing", and are not thinking about the interaction between their own trades and those of trade hedgers. That's fair enough, because neither are trade hedgers thinking about how their trading may or may not interact with or depend on the activity of non-trade operators. But there is one thing that is rarely considered: if the LME did not exist, the trade would still be mining, smelting and processing physical metal. And if funds or any other speculative interest decided an investment in metal was a good idea, they would still make that investment, LME or no LME. One can only guess at the mayhem it would wreak in the market if the funds moved into a physical market with the sort of money they have shown they are able to deploy on the LME!

CHAPTER 6

Legitimate fears

So far this book has been directed towards dealing with fears of operating on the LME which arise from ignorance or insufficient knowledge of the reality of the market as a hedge tool in particular. Now, however, it is time to deal with more precise fears about risks that could be associated with operating on the Exchange. Many of these fears are encapsulated in the sort of questions any businessman should ask himself before, for example, opening a letter of credit:

How dependent am I on the creditworthiness of my business partner?
What is the creditworthiness of my business partner?
What is the creditworthiness of those that my business partner does business with?
What framework of law, regulation and business practice lies behind the contract I am asked to sign?
What happens about force majeure? Or delayed shipment?
Can I get a nasty surprise from an import duty or turnover tax?
What assurance have I the material will meet contract specification?
Could the LME itself interfere with the completion of one of my contracts?

The totality of the answers to the above questions are necessarily spread through this and the next two chapters, but in this chapter we will outline the answers to the above questions.

CLIENT/BROKER RISK

The creditworthiness risk as between client and broker is fundamentally no different from that between any two business principals. However, whereas the broker has no additional assurances about the client beyond normal commercial checks, the client does have the comfort of the broker's need to meet capital adequacy rules that form part of the regulatory framework within which he operates as an authorised broker on a regulated exchange. If the client is a trade client whose interest is primarily in hedging, it is likely he will want the broker to extend credit to him from time to time to avoid the need for irritating margin calls as a hedge transaction on the LME winds its way to maturity. In the light of the abovementioned dependence of the broker on normal commercial credit ratings to assess the client, it is hardly surprising brokers' checks on new clients in this direction are extremely thorough, and would certainly include seeking bankers' references and probably also guarantees.

BROKER/BROKER RISK

Since every dealer in the LME Ring must deal with every other ring dealer who is bidding or offering at the market, there is a theoretical multilateral, if not systemic, risk if any one dealer were to fail. However, this has been taken care of since 1987 by virtue of the exchange joining the rest of the commodity futures industry and becoming a cleared market. This means that responsibility for assuring the execution of all contracts concluded across the Ring is transferred from the brokers who are parties to the transaction to the Exchange's clearers. These are the London Clearing House (LCH), a professional international clearing service house owned by six UK clearing banks. It has a capital of £15m and is supported for its performance by shareholder bank guarantees of £150m. The responsibility for completing the execution of trades across the Ring is transferred from the brokers to the LCH by what is called novation.

Because the LCH is providing essentially a financial safety net, it does not in fact undertake to complete the transaction exactly as initiated. It would not normally, for example, deliver metal in satisfaction of a sale, even though that might be what the buying broker's client would prefer. It will, however, assure that the buying broker will not suffer financially from a failure of his business partner across the ring. This in turn ensures that his client does not suffer financially. Under all normal circumstances, closing out all the open positions of a defaulting broker should not involve huge sums of money if all the contracts have been properly set up and margined. However, it is precisely because it is likely to be abnormal circumstances that might precipitate the failure of an LME broker that the LCH's resources are so substantial. It should never be forgotten that the LME suffered one of the biggest client defaults in commodity market history in 1985 when the International Tin Agreement, an inter-governmental undertaking, collapsed with debts of £900m, much of which was owed to LME brokers. Nevertheless through firm action by the Exchange the crisis was resolved and there was no systemic failure; and all this was at a time when the Exchange was still a principals' market.

LEGAL AND REGULATORY FRAMEWORK

Most of the data relevant to this heading is contained in chapters 7 and 8. But this is a good point at which to draw attention to the LME's arbitration service. This is sufficiently well-known and respected in the international metal trade that it is quite common for contracts that have nothing to do with LME trading to include a clause providing that any point in the contract requiring arbitration shall be settled by LME arbitration. When there are many internationally-respected alternative fora for an arbitration hearing, this preference for LME arbitration is one more reflection of the LME's close connection to the realities of the physical markets that underlie its futures trading.

FORCE MAJEURE AND DELAYS

It is a central principle of LME contracts that there is no such thing as force majeure. With a commodity futures market dealing in standardised quantities and qualities of metal, it is always possible to buy in other warrants to replace those that the vendor intended to deliver, but found he couldn't. Equally, a buyer offered delivery in satisfaction of a contract who has no means of taking up the metal can sell it on in the market. As we have seen, by far the majority of hedge transactions conclude by buying back or selling the futures position that has been established. The question of force majeure does not, of course, arise with Asian options, which are always cash-settled. All of the foregoing gave rise many years ago to the rather hard little ditty:

> He that sells what isn't his'n
> Buys it back or goes to prison

This thought should not unduly deter short hedgers, as they of course intend to buy back their short position at the conclusion of the hedge anyway.

The whole machinery of trading only in warrants on metal which is safely in LME warehouse is primarily geared to supporting the principle of "no force majeure".

DUTIES AND TAXES

An international organisation like the London Metal Exchange is nearly always struggling with some one or another piece of strictly national legislation framed by parliamentarians who cannot believe that there is a world outside the electorate who vote for them. In its time, the LME has suffered from – and survived – UK legislation on import duties and currency convertibility which had tremendous potential to distort or frustrate its trading activity. The central principle behind all LME contracts is that the metal shall be traded duty-free, as though it were a cif contract, even though it is in fact in-warehouse. This is easily achieved where LME warehouses are in free port zones, or in countries where the particular metal is not subject to import or export duty. It is not so easily achieved in other circumstances, and this is where the LME's ceaseless vigilance in defence of its trading practices is needed. It is also one of the reasons that not all LME-registered warehouses store all LME metals. If the problems of import duty or tax treatment of a particular metal in a particular country have so far proved insuperable, then there is little point authorising warehouses for that metal in that country.

Nowadays, of course, it is not only national legislation that can give problems. For much of the nineties the LME has been patiently negotiating with Customs authorities in various European countries on special exemptions needed for LME-traded metal delivered to buyers in that country when it comes to liability for value-added tax (VAT, tva, mehrwertsteuer or whatever). Obviously, if metal is subject to different tax burdens upon delivery out of LME warehouse, depending on where the person to whom it is being delivered is located, this will have a distorting effect on trading across the Ring. The LME authorities were well-equipped to enter these negotiations, having debated a broadly similar issue with UK Customs and Excise a decade earlier, when there was a risk of VAT being payable on every single tonne of metal traded across the floor of the Exchange. That problem was resolved by securing the agreement of Customs to view LME trading as occurring within a "black box" where normal VAT rules did not apply.

As this book goes to press, it has been announced that the tricky question of VAT in the countries of the EU on metals traded in-LME-warehouse has been resolved. Under new EU legislation, both EU- and non-EU-origin metals can be traded free of VAT in-LME-warehouse, though the dates of effectiveness of this legislation may be as late as Jan 1 1997 in the case of Germany and Luxembourg. The new Directive will thus progressively supersede the series of bilateral agreements between the Exchange and member states which had already abated some of the practical pain of this issue.

Broadly, therefore, the answer to this question is that there can be duty or tax angles to physical delivery of certain metals in certain countries, but these are all known and recorded. So at least there should be no nasty surprises. What is more important in broad terms, however, is that the LME has worked, and continues to work strenuously to ensure that these occasional abnormalities do not react back on trading across the Ring.

CONTRACT QUALITY

Sooner or later in the world of futures markets, readers will come across the expression "sardine can trading". This refers to the risk that a standard commodity will become so familiar as a trading token that it ceases to be useful as the commodity it is supposed to be – in other words, a can of sardines changes hands dozens of times, but when it is eventually opened, the sardines are rancid. The image is kept refreshed by various commodity trading scandals that have occurred in the past – frozen orange juice and grain having provided some notable examples. In general, metals are less susceptible to these problems as they are of identifiable provenance and do not deteriorate in storage. However, as new contracts have been added in more recent years, there are some exceptions to this general rule.

Nickel is indeed traded on the LME packed in drums, and a newcomer to LME trading could be forgiven for experiencing a frisson of doubt that the drums delivered to his works as nickel in fact contained a solid lump of iron. However, a glance at the nickel contract reproduced at the back of this book will show that the LME was well aware of the perils of sardine can trading; anyone placing nickel on warrant must expect to pay for the warehouse to empty the drum, inspect the contents and re-seal or re-drum under its own seals unless the metal is warranted by a Ring dealer. As to the nickel itself, it is the branded production of a major producer who is answerable both to the LME and to the court of its reputation for the quality of its metal.

Aluminium alloy is another exception to the rule that metals in storage, in general, take care of themselves. Its known tendency to deteriorate in storage, at least as to surface appearance, is catered for by the twin requirements that warehouses shall store it in appropriate humidity-controlled conditions and that metal that has been taken off warrant may not be put back on again. Funds, as well as a few other Exchange users, are likely to be put off trading this contract because of this requirement. Again, the metal chemistry is assured by its branding, and even by melt number certificates, which must accompany the batch.

An indication of the effectiveness of the LME's systems for policing metal on warrant is given by the movement of primary aluminium stocks in 1994/5. During the peak pressure on warehouses in 1993/4, the LME introduced concessions for the storage of primary aluminium in outside compounds under tarpaulins. As a million tonnes came out of storage into consumption as the market began to level off, there were no reports of quality disputes.

THE LME AS "BIG BROTHER"

Study of the LME's Rules will show that in certain circumstances the Exchange has quite wide powers to intervene in the normally fairly unfettered process of day to day trading on its market. These powers are reserved for situations of impending distortion of, or danger to the market. In the past decade or so (a period which just includes the notorious Tin Agreement collapse), these powers have been invoked on some ten occasions.

Since the introduction of large position reporting, the Exchange can examine the transactions of ring dealers with particular open positions in the metal, or for the date, to which anxiety attaches. Usually the trigger for such action is the build-up of positions which might indicate an attempt at a squeeze or even a corner. As the account of the history of the Exchange at the end of this book shows, a desire to corner one or other of the LME's metals keeps recurring, notwithstanding the fact that no attempted corner has ever really succeeded. Symptoms of such a move are often the emergence of an abnormal backwardation on a particular date or dates. Because of the importance of options declaration and settlement days in the month, it is often hard to tell whether one is seeing a normal, temporary and not too serious blip in prices associated with one of these days, or whether someone is using the cover of such a likely move to build a more sinister position.

If the Exchange is still not happy after examining dealers' cards (which, of course, is done in the strictest confidentiality), it may impose limits on the daily backwardation at the sensitive time. At this point, there will almost certainly be a double squeal from the market. Those who were, deliberately or by chance, going to benefit from a sharp blip in the backwardation claim they have been robbed, while those who are crunching through a routine hedging programme and will be inconvenienced by any abnormal level of backwardation claim the Exchange is not doing enough to protect them. If the two squeals are of about equal volume, the Exchange has probably called the limited backwardation about right.

Strongest of all the powers of intervention available to the Exchange is its ability to establish a "ring-out". A known device in the world of futures markets, a ring-out on the LME had never been seen before the collapse of the International Tin Agreement precipitated a whole series of unprecedented events. The LME was thus obliged first to give itself the power to introduce a ring-out, and then to operate it. In essence, a ring-out is associated with an event so serious as to call for the temporary, or even permanent, cessation of trading in that contract. For this to happen, a figure must be nominated at which all open positions are closed, and it is nominating that figure which constitutes a ring-out. The Exchange was, at that time, challenged in the courts over both its assumption of ring-out powers and its manner of exercising them. The Court decided it was right on both counts, and today the power to impose a ring-out in appropriately extreme circumstances is enshrined in LME rules.

For the ordinary user of the Exchange, especially the trade hedger with a continuing programme of transactions, the idea of the music suddenly stopping with a ring-out is even more distasteful than a severe squeeze. So it is important to get all these things in proportion. The collapse of the International Tin Agreement was not only the most dramatic event in the Exchange's then 109-year history, it was also 96 years after the second most infamous event in the life of the Exchange, the Secretan Syndicate. Since the lessons of each exceptional event of this type are learned, and fresh armoury put in place to prevent a repetition, the reader may calculate for himself the odds against another Secretan or Tin Agreement in the next hundred years.

The principle which all this establishes is that the risk-avoiding trade hedger will regard an impending abnormality in the curve of prices as yet another hazard to be avoided, and will listen carefully if his broker alerts him to such a hazard. He will leave it to those of a more avowedly speculative intent to consider whether to take the LME's equivalent of a white water ride by plunging into the impending turmoil.

CHAPTER 7

Nuts and bolts

Total professionalism in trading on the LME requires the broker to know the LME rules and regulations inside out, and the client to have at least a good working knowledge of them. This chapter has therefore been designed to be used as a "one stop" reference for those who already have this book on their desks. It must be stressed, however, that there is no substitute for the real thing – ie referring to a correctly-updated LME Rules and Regulations book.

We will set the scene by noting the corporate environment in which the LME works. London Metal Exchange Ltd, the company responsible for controlling and maintaining the operating framework for the Exchange, is a private company limited by guarantee, that guarantee being £100 for each member of the Exchange. It would be called in the event of the company becoming insolvent and members are liable even for twelve months after ceasing to be a member. As a "capitalisation" this gives no idea of the scale of the Exchange's business as an exchange. In fact, the 1994 Annual Report showed earnings of over $250 million from a trading volume of some $2,000 billion. LME Ltd's revenue stream comes from entrance and membership fees and a contract levy, and it has liquid resources at any one time equal to six months' turnover. It was sufficiently active and successful in recent years that in 1994 it was possible to abate the scale of the contract levy, though this could be reversed at any time it was necessary.

Like any company, LME Ltd is run by a Board whose composition is prescribed by the Company's Articles. Of a total of 16 members, eleven are elected from the Exchange membership, four are invited and the sixteenth is the Exchange's chief executive for the time being. Of the eleven directors elected from the Exchange membership, eight, or 50% of the total Board, are from Ring members, the other three coming from Associate member firms or the individual membership. These directors are elected for three year terms and may not serve for more than two consecutive terms, though they are free to stand again after a year out of office. Originally there was an idea that of the four invited members one or two should have no connection with the international metal trade but come from "the great and the good" for reasons explained in the next chapter. In the event, these places have much more constructively gone to individuals who represent various sectors of industry which are inherently Exchange users. They are appointed on the basis of annual re-election and may not serve for more than twelve terms consecutively.

Voting on the Board is somewhat complex. In voting on elections to the Board, each member has one vote. On all other matters, however, votes are weighted with five for ring members, three for Associate clearing members and one for Associate members. Thus on voting the Ring members can easily muster a majority.

INFLUENTIAL COMMITTEES

As an inevitable effect of the Exchange becoming a cleared market and of the bureaucratic requirements of the Financial Services Act, the scope and at times heavy responsibilities of the Secretariat would astonish anyone who had become used to the very modest administrative support considered necessary for the old principals' market. Where there is a legacy of the old Board and Committee structure (as detailed in Chapter 19 of this book), is in the range of administrative committees, all made up of volunteers from the Exchange membership, which manage many specific aspects of Exchange business and maintain a two-way dialogue on their important issues with the Board. They cover such matters as contracts (one committee for each metal), arbitration, warehouses, trading discipline, options and publicity. Like the Board, they can count on substantial support from paid Secretariat staff.

Compared with the dichotomy of the Board and Committee structure previously, the present Board encapsulates in one body the legal framework of a Board, the democratic representation of the Exchange membership (previously confined to the Committee), the balanced representation of brokers and clients, and, with the stronger Secretariat, a better balance between professional support and volunteer input to the work of the Exchange from broker and client members. However, as the continued process of revising the rules reflects, no institution has a permanent shape, nor does it operate in a stable environment; the situation of which a snapshot is given here will undoubtedly continue changing over time.

MEMBERSHIP

Corporate membership of the LME is divided into five classes which are permutations on the two broad splits between clearing and non-clearing and between broker and trade. The names and addresses of all the companies concerned as at the time of going to press with this edition are listed in Appendix IV. Most select of these groups is the Ring dealers who are the only class authorised to deal in the Ring; they are therefore by definition brokers and of necessity clearing members. Associate broker clearing members have all the same rights and obligations as Ring dealers except for actually dealing in the Ring. Associate trade clearing members have the rights and obligations of clearers, but obviously do not deal in the Ring and, because they do not deal as brokers on behalf of others, do not need to be subject to the many provisions of the FSA that apply to brokers.

Non-clearing associate broker members is a self-explanatory group incorporating many of the introducing brokers that are active on the Exchange, while non-clearing trade members are smaller trade clients of the Exchange who are sufficiently active they wish to benefit from being part of the Exchange population. Last but not least are the individual members who, as their title suggests, have no corporate affiliation, but are active on or through the Exchange and who may, if they so wish, stand for Board membership. When it comes to negotiating with one's broker over commissions, one's status as an Exchange member is one of quite a multitude of factors that may be relevant.

THE TRADING PROCESS

One of the first things a broker is likely to say to a prospective client is "I'll take you down to the Ring and you can see for yourself what goes on". (Security considerations put a stop to this practice and to public visits to the visitors' gallery for a period of years when the Exchange was in Plantation House, but its normal posture is one of openness). However, a visit to the Exchange does not necessarily give a visitor the same kick it gives the broker who is taking him there. It may well appear to him that the Ring is completely out of control, that nobody can possibly hear what anybody else is saying, and that even if they could, nobody could possibly prove afterwards that they had said it!

In practice, regular attendance at the Ring for a period of weeks allows anybody, even someone who initially thought he could never grasp any part of what is happening, to familiarise himself with the proceedings and to report accurately on where prices have been and who traded with whom. The fact that member firms have their allotted permanent seats around the Ring, unlike the shifting sands of a pit, helps in this regard, as does the LME's practice of dealing in one metal at a time in short Ring sessions. The one proviso is that such an individual should have normal eyesight and hearing and preferably be aged under 30! With the division of labour in brokerage houses nowadays between ring personnel, client liaison staff and the back office, it is in fact not necessary for the client to have such knowledge of ring dealing that he could, at a pinch, do it himself. He can rely on the Exchange's Rule Book, the SIB, the FSA and inter-broker competition to ensure that Ring dealing is conducted as reliably and honestly as he would wish it to be. What he has to work on is the productiveness of his relations with his client liaison contact.

Some key rules are often referred to by the effete name of "trading etiquette", but their stringency, and the severity of the penalties for breaking them, should not be underestimated. All bids and offers must be audibly addressed to the whole Ring, and any broker can deal with any other at his shouted bid or offer price. Assuming no quantity is specified in conjunction with the bid or offer, the broker who is willing to accept or pay the quoted price will normally indicate the number of warrants he or she wishes to deal at that price. More likely, he or she will test the water by vociferously counter-offering and bidding in an attempt to influence the trading range. Sooner or later a bidder's or offerer's price will be accepted.

If an initial offer is accepted and a significant number of lots are taken up, either by the first bidder or by several bidders, the price for that position will move up. If the first bidder is also keen to see the market rise, he may enquire of the offerer "How much?", a dealing convention to indicate that he is in the market for 20 lots at that price. The offerer can deal at that level and, if he has more to sell at that price, resume offering. Or he can reply "No, only 5 lots". At this point other brokers who were, or are now, offering at the same price will also deal with the bidder. A variant on the above situation is for a broker to respond "Yes" to a price which is bid or offered. This means he will deal in 20 lots with the broker to whom he has responded, but is not making an open commitment to the whole Ring. It is normal for dealing on this scale to be concentrated in relatively few active dates – cash, three months, a few Wednesdays in the further forward positions and perhaps a couple of dates sensitive to an impending squeeze.

In this connection it is worth recalling that LME trading dates fall on every business day in the first three months of trading. Normally, the only days that are not business days are UK or US Bank Holidays, but the LME has the power to declare any day not to be a business day if there is a sufficiently strong and abnormal reason to do so. In the fourth, fifth and sixth months from cash, every Wednesday in the month is a settlement day. Thereafter, ie from the seventh month to the limit of trading in the relevant metal, settlement day is the third Wednesday in the month, the same day as option declaration day. In the case of aluminium, copper, nickel and zinc, the limit of trading is the 27th month following cash, while for aluminium alloy, lead and tin it is the 15th month.

Settlement itself is a feature of all futures markets and need not really concern the client as his execution and final statement on a deal would be the same however settlement was effected. In fact, settlement is a way of short-circuiting what would otherwise be an unwieldy number of bilateral transactions, involving huge sums of money, exactly reflecting the dealings done between the respective brokers. It works by expressing all the deals done that day by a particular broker in terms of their margin of difference from the settlement price. Each broker then pays into the market, or receives from the market, only his total difference from the settlement price for each metal according to his actual contract prices and tonnages traded.

Likewise, where contracts are settled by delivery of the metal (in the form of the relevant warehouse warrant), settlement avoids the "daisy chain" of dozens of back to back contracts familiar to anyone who has dealt with, for example, physical oil trading. The broker who has a net obligation to deliver warrants to the market does so, and they are received by a broker who has a net obligation to take up warrants on that metal from the market, even though these two brokers may not have traded that metal between each other at all that day.

A key input to the clearing house is the official notification from the LME of the official prices which are read out each day almost immediately following the close of the first session. Established by a quotations committee of three senior exchange members (each committee is on duty for a week), the prices cover cash, three months, fifteen months and 27 months. The quotations are susceptible to immediate challenge from the Ring under a formal procedure, and if the committee agrees with the dealer's presentation of the situation an amendment will immediately be announced. The quotations committee then does it all again after the afternoon session, though as the prices are unofficial they are not read out. They are, however, input to the clearing house as closing prices, together with the exchange rates that are to be officially used for settlements in the major currencies other than the contract currency (US$). These are £ sterling, DM and Japanese Yen.

As this book goes to press, it has been announced that the LME's 100+ years of calling the official prices after the morning session is to give way to an electronic equivalent whereby provisional closing prices, established by the quotations committee, will be flashed up shortly after the second ring of each metal. If this close has been unanimously agreed by the independent judgment of each member of the committee in all positions within five minutes, it will be displayed electronically in green. If, after five minutes of display, it has not been contested it will be displayed in red and will be official. This will avoid the long lag between the close of the early rings (copper, tin and lead) and the official call, as at present. In the case of any disagreement about the prices, however, the provisional prices will not be made official until the quotations committee has deliberated again after the session.

A GOOD MATCH

In a global business like metal futures, a day is a long time. Lights will be on in many London brokers' offices from 5am or 6am (to service Far Eastern clients and keep abreast of those markets) until 10pm, 11pm or midnight, which is still slightly short of the end of the business day on the US West coast. Despite the central importance, both operationally and in terms of market philosophy, of the LME's two Ring trading sessions each day, all deals done on an official basis between clearing members have to be cleared.

For the clearing house the trading day ends at midnight, and any deals done too close to that deadline to be entered and matched are carried over to the next day. Each day is broken up into matching periods to give the clearing house a chance to keep on top of what is going on among the thousands and thousands of bargains that are continuously being struck.

Clearing members are obliged to enter all transactions into the computer system that feeds data to the clearing house. Once matched and accepted by the clearing house, the deal is then assured of being settled. The next day each clearer is able to access, over the same computer network, advice as to what cheques he has to make out to, or receive from, whom and what warrants he has to deliver to, or receive from, whom.

Additionally, he receives information on his own overall position in each of the markets, including the amounts he has to remit as initial and variation margin on all his various trades. These amounts he will partly fund from his own resources where he is able and willing to extend credit to his trade clients for such purposes. But for clients who are required under the FSA to have, or who have opted to have, segregated accounts, the broker will call the client immediately for these margins. This knowledge of his own overall position is, of course, vital for the broker to be sure he is keeping his exposure within his capital adequacy limits, while detail on each client's evolving position is equally important in terms of his credit limits. Last but not least of the information the broker can obtain from the clearing house is a statement of his global position and exposure on all the futures markets cleared by the LCH; these include markets in financial as well as commodity derivatives. In terms of meeting capital adequacy requirements, it is his global position over all these markets that counts, together with any OTC business he has executed and reported to the SFA, but which does not go through the clearing.

LME CONTROLS

It is not only the broker himself who can access this data from the clearing house; the LME itself has powers to inspect the position of any clearer or any client at any time. This may be to check the position of that broker or client alone; or it may be to detect an unusual build-up of trades between a pair of brokers or clients, or a few; or to detect at an early stage an abnormal concentration of activity on a particular date, which could be the precursor to a squeeze or corner. Additionally, brokers are obliged to highlight for the notice of the LME and the regulators any large positions being built up by any client. Reacting, albeit a little belatedly, to the lessons of the collapse of the Tin Agreement, the Exchange's powers of investigation and intervention are now considerable. Moreover its rules now incorporate a disclaimer of liability for the Board and its members while exercising these powers which should preclude a repetition of the litigation that was attempted in the wake of the Tin Agreement collapse.

Also linked to that event, as well as to more recent upheavals involving Metallgesellschaft and Codelco, extensive new rules have also been introduced to deal with what might be called peri-LME trading. Described in the Rules as Non-Market Price Transactions, these involve principal-to-principal transactions, probably tailored by an LME broker or an introducing broker to meet a particular need of a client. They probably also depend on being backed up by a futures market transaction; but in that they are not cleared transactions in themselves they carry several elements of risk, notably that of escaping the scrutiny of the prudent and regulatory oversight of the clearing house and the Exchange. In accordance with normal LME practice, especially where trade-oriented deals are concerned, the new rules do not ban such transactions outright, but they do impose stringent criteria on such matters as written authorisations and audit trails.

Further rules have also been introduced covering transfers of responsibility for cleared open positions from one broker to another; these are often erroneously referred to as "give-ups". This could arise when a client might have, for example, an open position with one broker which is in profit, whereas his overall position with another broker is perilously close to his credit limit. Moving the profitable position from the first broker to the second would not precipitate a problem with his overall position with that broker, but would avert possible margin calls from the second. In these circumstances the client might instruct the first broker to transfer the position to the second, who would then assume responsibility eventually for liquidating it.

Strictly speaking, give-ups involve only trades which have not yet been entered into the clearing. A client might, for example, instruct a broker to execute an order for him because he favours that broker for executions. However, if he has no credit line with that broker, he would then want the executing broker to give up the transaction to his clearing broker to be entered into the clearing and eventually settled.

COMMISSIONS

Wherever brokerage is concerned, the broker's commission can become a hot topic. In absolute terms, LME commissions are modest compared with those prevailing in many other branches of financial services. Commission on a "round turn" (ie purchase and sale, or vice versa) will normally lie somewhere around 01.25-0.25% of contract value, payable half on the first leg of the transaction and half on the second. When it comes to options, the broker, especially if he is a ring dealing broker, will most likely be the market maker himself and will sell the option outright to the client. Compared with differences, margins and the outright cost of a warrant of metal which the client takes up, commissions are a fairly modest item on the client account.

CLIENT PROCEDURES

The next chapter explains the background to today's formal and rather bureaucratic requirements for "vetting" a prospective client which a broker must undertake. Under the "know your client" requirements of the Financial Services Act, clients must be asked for information that allows them to be classified as "Ordinary business investor", "Market counterparty", "Trade customer" or "Expert customer" on the one hand or "Private clients" on the other. The requirements are not particularly onerous.

A trade customer has only to demonstrate that it uses the futures markets as an integral part of its business.

A market counterparty has to show membership of the SFA or a futures exchange, or be an overseas investment business.

An ordinary business investor may be large (with net assets of £5m or equivalent or net assets of over £0.5m and 20 or more shareholders) or small if it classifies as a trade customer.

An expert investor may be individual, joint venture or corporate. They have to show that an individual, in the case of individuals or joint accounts, or the directors and senior management in corporations, have experience and understanding of futures trading.

Needless to say, the foregoing are the types of clients most LME brokers are looking for. Larger houses like Rudolf Wolff do not accept private clients directly, though they of course service them indirectly if their business comes through a managed futures fund or hedge fund. Private clients, who may also be either individual or corporate, are those who fail to match any of the above classifications.

As a result of knowing his client, the broker assigns each one a position limit. Unlike a credit limit, which is a simple cash amount, a position limit is built up taking cognisance of the assessed risk of each transaction, which is also expressed through the size of the initial margin. All this is built into the client's account records on the firm's computer which can then signal if the client is in danger of overtrading or breaching his credit limit. The account executive then begins the process of maintaining contact with the client, discussing his trading strategy with him and taking his orders to trade.

Learning to live with an Account Executive is rather like learning to live with a computer – you must at all times remember that everything needs to be specified. The client may be quite certain his conversation with the AE has been all about copper, so when he says "OK we'll buy five lots three days over cash", he may feel it quite superfluous to specify that copper is the metal he wants to buy. It isn't! The AE might have his own equally good reasons for believing the conversation was all about aluminium, or he could be in the middle of five people screaming their heads off about aluminium. Of course he should be sufficiently in tune with the client's thinking to get it right even if the metal is not specified, but there's always a risk that he won't. The effort involved in saying "Buy five lots of copper three days over cash up to $2950, good till cancelled" is scarcely more than saying "OK we'll buy five lots three days over cash", but it is many times more efficient in getting the exact message across. Other criteria that may be central to the purpose of an order could refer to the close of a ring, a specific stop-loss limit, or even be contingent on another event such as liquidating an open position first.

DOING THE BUSINESS

The AE has now taken the client's order down on a time-stamped order slip and it is conveyed to the broker or Ring dealer who is responsible for executing it (if possible). A sudden market movement may make the order completely non-feasible in the foreseeable future, and in this case the AE needs to go back to the client and explain what is happening. More probably, he will be able to report "mission accomplished" in a few minutes, by which time the trade, complete with the all-important price at which it was done, will be logged in the firm's computer ready for the preparation of the next client statement. It will also, as a by-product of the transaction, enter the clearing house computer for matching.

Next morning, the AE will have confirmation of that trade (and all the other trades the client has ordered that day), together with an overall client position review showing his open positions and how far these are in profit or loss on a mark-to-market basis. A further by-product of the computer processing is the actual client contract, which will be sent to him as further confirmation. All this data rolls forward into the client statement at the end of the month unless, meantime, a movement of the market puts the client account outside position limits, when a quick margin call on the client may become necessary.

MARGINS EXPLAINED

It should be explained that it is the very nature of futures trading, and the fact that virtually all trades are done with the leverage of margin right up to the point of settlement, that makes it so important to keep initial margin in tune with the ebb and flow of the position's gains and losses. Whether the margin is provided by the client himself (and on US exchanges it always is) or by the broker on behalf of the client, it has at all times to represent the same proportion of the total value of the contract that was deemed prudent at the outset. If the initial margin was, for the sake of argument, 10% of the contract value and the market then moves 10% against the open position, the variation call is not 10% of the initial margin, but more like 100% of it.

An example will illustrate this. Suppose a client bought ten lots of aluminium on margin at $2000 per tonne, that is a contract value of $500,000 and the initial 10% margin is $50,000. The market falls by 10% (highly unlikely in one day, but it illustrates the point). The contract value is now $450,000 and the loss on the client's open position is $50,000. It is this loss which the variation margin call is primarily interested in covering, though there is a small offset in that the initial margin of $50,000 is now $5,000 more than is required to maintain 10% of the contract value. The client would therefore be called for $45,000 and this would be payable within 24 hours. It could be provided as cash (most major currencies are accepted), bank guarantee or other acceptable security, such as metal warrants held by the client, though these would not be credited at 100% of their face value. An offer of power boats, works of art or property would be unlikely to be acceptable! If the futures contract on which margin is being called is a straight hedge transaction, the client's only problem is one of cashflow as he will, of course, be enjoying an equal and opposite profit on his physical position.

As already explained in outline, on the LME, favourable margin on an open position in profit is not paid out until the contract reaches maturity. On US markets, it generally is. In LME accounting, accrued favourable margin serves to strengthen the client's open position, but it does not accrue interest. Only cash in the hands of the clearing house or broker can earn interest.

This distinction between where the open position stands and cashflow is particularly important for the question of credit control. Most LME brokers extend credit to trade hedge clients to cover initial and variation margin and their limit on this credit is important both for the client (as a subsidiary control on overtrading) and for the broker (who has his own overall capital adequacy limits to respect). As with margin call, therefore, any market movement that puts the client in breach of his credit limits will precipitate an immediate call for more security. Again, this can take the form of any acceptable major currency, bank guarantee or other acceptable security.

CHAPTER 8

Regulation

Regulation of commodity markets in the UK is nothing new. As far back as the Middle Ages, markets and itinerant traders were regulated under special laws that operated in parallel with common law. Whether it was the impressively-titled Lex Mercatoria or the more engagingly-titled Courts of Pie Powder, there was a framework of regulation peculiar to these markets and traders. This was needed to resolve commercial disputes arising in an environment with which common law was not much concerned. (Students of history can read the whole story of regulated commerce from Roman times to the present day in a book entitled "From Forum to Futures", published by Futures & Options World).

However, readers of this book will be much more concerned with the present-day framework of regulation which has evolved over this long period of history and its impact on their conduct of business with the LME today. To explain this unfortunately requires the use of a number of acronyms.

Modern regulatory history in the UK begins with the 1986 Financial Services Act (FSA), which created as its principal regulatory body the Securities and Investments Board (SIB). The SIB has a two-pronged attack on regulation – via the Exchanges and via individual operators in the financial services field, grouped according to their main interests. On the exchanges side of the operation, SIB recognises exchanges as capable of being relied on to maintain standards by self-regulation. These are called Recognised Investment Exchanges (RIE), and the LME is one such. Smaller than the London Stock Exchange (LSE) or Liffe, LME is more heavyweight than the International Petroleum Exchange (IPE) or the London Commodity Exchange (LCE) which trades softs. SIB is also responsible for recognising and monitoring clearing houses where these are separate from the RIEs, as is the case with the LME.

On the operators' side, various trade authorities are also recognised as capable of self-regulation of their members, and they are called Self-Regulatory Organisations (SRO). Originally, the SRO to which LME members belonged was, quite understandably, called the Association of Futures Brokers and Dealers (AFBD – four letters for a change). However, after a number of years, and not least in the wake of the need for the London Stock Exchange's own options market LTOM to merge into the major financial futures market in London, Liffe, the AFBD was merged with the Stock Exchange's trade association to form the Securities and Futures Authority (SFA – not to be confused with the FSA). The responsibilities of an SRO can be summarised as follows:

– to make rules for its members, based on SIB's principles
– to vet membership applications and authorise members
– to vet, register and test individuals
– to oversee members' conduct of business
– to monitor advertisements and promotional material
– to monitor the security of client money held by members
– to ensure members maintain adequate financial resources
– to enforce rules and penalise breaches of them
– to deal with complaints and arbitrate disputes

ARE YOU AUNT AGATHA?

Other groups of financial services companies dealing in insurance, pensions, portfolio management, etc. were covered by further organisations known as Fimbra, Lautro, Imro and so forth. Just as everybody was beginning to settle down to this complex framework, the powers that be decided it was time for a change. From the outset, the FSA has been accused of an excessive preoccupation with "Aunt Agatha", the proverbial amateur, inexpert investor liable to take misleading advice on investment into stocks and shares, insurance policies or funds. Obviously, such a person is a far cry from the typical client of a commodity futures market, and especially a trade-oriented market like the LME. Unfortunately, whilst every legislator and civil servant concerned with the framing of the FSA could imagine himself an expert on Aunt Agatha, very few had any idea what went on on commodity futures markets, or indeed derivatives markets generally.

Thus despite hours of representations by the professional exchanges that their regular business would be seriously hindered by a law which assumed their clients were somebody they actually weren't, and despite a large number of modifications to the original drafts, there is still an element of bureaucracy about the FSA which is irrelevant to, and counter-productive for, exchanges like the LME. Now the champions of Aunt Agatha have decided that the SIB and the SROs are still too professional for their liking, and a new body is being formed called the Personal Investment Authority (PIA). This is to slot in alongside the SIB and separately – and directly – control members of present SROs like Fimbra and Lautro whose clientele is more or less in the Aunt Agatha mould. Fortunately, this will be almost entirely irrelevant to LME brokers and their clients, as Aunt Agathas are very few and far between among LME clients.

However, another more laudable principle of the FSA is that brokers should "know their client", and this is why a new client must complete paperwork telling the broker about himself and his trading objectives on the LME, even if the two have in fact known each other for years. Yet more paperwork is designed to trap the occasional drug baron who is using the futures markets for money-laundering purposes, so again, the client must not mind if his broker appears to think he is a drug dealer. The foregoing should be sufficient for a client to establish that he is not an "Aunt Agatha".

SEGREGATION AND BEST EXECUTION

A helpful and practical consequence of a client being duly confirmed as having trade or professional status is that he is entitled to opt out of the provisions in the FSA for client funds to be held in a segregated account. At first sight, it might appear suitable for a trade client, just as much as is obviously the case for a private client, to have a segregated account. On the LME, however, the corollary is that he would then have to accept full margining on all his positions, rather than benefiting from the credit that LME brokers expect to extend to trade clients to save them both the trouble of raising myriad cheques over the life of a contract. Obviously, the clients of a broker will be due margin payments in both directions, so the amounts of credit he is extending to clients do not necessarily add up to huge sums. In any case, his own exposure limits and capital adequacy requirements under LME and SFA rules ensure that he cannot, even inadvertently, over-extend himself in this way.

Another facility which is contained in the FSA in the spirit of looking after Aunt Agatha is a requirement for a broker to give his client best execution on his order. In the rapid cut and thrust of an open outcry ring, this is a counsel of perfection which is simply not attainable in practical terms. If a broker has a buy order at, say, \$1700 for ten lots; if the market is in a strong upsurge; and if the most that is ever offered in that particular ring at that price is two lots, then within the meaning of the act, he cannot achieve best execution. In today's unfortunately more litigious climate, promising a trade client best execution on a market like the LME would be close to suicide. Not surprisingly, trade clients of LME brokers sensibly accept that if their broker is to act for them to his realistic best, they have to exercise their option, under the FSA, to forego the right to best execution. Once again, the sanction of losing a client if he is not satisfied with his executions is a more direct spur to the broker doing his best than any clause in the FSA. Equally, the broker has to be careful to handle his client's business strictly in accordance with the Act, as to do so otherwise is to commit a criminal act, and also to create a contract which is not legally enforceable.

To be accepted as a member of the SFA, an LME brokerage firm must satisfy the Authority it is "fit and proper" to undertake brokerage activity in which a high degree of trustworthiness is essential, not only in handling high value transactions and securities (warrants), but also in executing client orders faithfully and without "churning" (committing the client to an unnecessary number of transactions in order to stoke up the brokerage bill). In a people-centered business like futures, a lot turns on the individual working for the brokerage house. So individuals, too, have to be assessed as "fit and proper". To this end, they are required to sit examinations to demonstrate professional competence, as well as to show a distinctly blameless cv. It will be remembered that in the notorious case of the Barings derivatives dealer Nick Leeson, he was not able to obtain a dealing licence in the UK simply because there was a record of a county court judgment against him. Front line personnel such as directors of brokerage houses, client liaison staff and salesmen have to pass examinations as Registered Representatives. Ring dealers and back office staff who do not come into direct contact with the client, but are vital to team performance, have to qualify as Registered Traders.

VITAL ROLE OF DOCUMENTATION

Operationally, the most pervasive effect of the whole FSA/SIB/SFA chain is the requirement on brokers to keep extensive documentation. When this relates to orders barked down a phone in the heat of a five-minute ring, it can be appreciated that a high degree of organisation is required, backed by all the assistance that modern technology can provide. This is evident on the surface of any visit to a modern dealer's trading room. Woe betide the dealer or firm whose paperwork does not balance exactly at the end of the day. They are bombarded with demands from the LCH and in short order the LME secretariat and the SFA to disentangle their book and show a spotlessly clean execution. Among all this, the firm's own compliance officer, too, will be reviewing the records to try to trace the mismatch.

Mismatches, of course, commonly turn up at the end of the day when one broker reports to the clearing house a sale to another broker which the latter does not record as a purchase (or vice versa). This in turn ripples through to the likelihood of a mismatch on client contract documentation as well. Where this occurs, it helps if everyone keeps calm and reflects that they are dealing with a tiny fractional percentage of all the trades concluded across the ring and in official inter-office dealing. At the end of the day, most mismatches are resolved with the support of the copious record-keeping required. Where they are not, it is usually the dealer who must go into the market as soon as practicable and deal on his own account to square the situation as best he can. In a fast-moving market, this may not be all that good.

Apart from nitty-gritty documentation, brokers must also provide documents relating to business conduct principles such as customer agreements and risk disclosure statements. Another thing the FSA is, rightly, hot on is brokerage firms dealing for their own account. Such trades must be particularly carefully logged, and where a client wants to deal at the same price, his order must at all times be given preference. In this connection, an important piece of required practice is the time-stamping of each order the second it is received.

Just as critical as the daily returns to the clearing house are the monthly returns each broker must make to the SFA covering financial resources, or as it now tends to be called, capital adequacy. This return must be made within three weeks of the end of the month, and if all is not in apple pie order, he can expect his licence to trade to be suspended. Such are the complexities of modern futures trading that there can still be debate as to how various asset classes or open positions should be counted in calculating financial resources, and from time to time case histories throw up new rules crystallising this aspect.

An important element connected to this issue is the granting of credit to clients. SFA guidelines on this matter are quite precise, and include the following:

- directors or officers who decide on credit should not be traders or salesmen

- the method of assessing credit and the amount individuals can grant must be stated in advance
- the customer must be notified in writing of the outcome of the above
- what is acceptable as collateral must be clearly defined
- appropriate client files must be set up
- credit lines must be reviewed annually in terms of both volume and finance
- client credit must be monitored daily
- clients must be advised of the steps that will be taken if they transgress the firm's credit rules
- clients must be advised the firm's policy on use of non-approved banks

REGULATION OF THE EXCHANGE

The regulation of the LME as a Recognised Investment Exchange runs in parallel with the regulation of member firms and their personnel and is no less strict. As with individual member firms, the Exchange is required to show sufficient financial resources to support its whole range of operations. To this end its reserves are maintained at at least six months' operating costs.

Most importantly, the Exchange must show that it conducts its markets in an orderly manner. Because of its relatively unusual system of running all contracts to maturity, rather than netting-off as soon as this is feasible, the LME has considerable powers under its rules to call for members' books to enable it (confidentially) to look for unusually large positions. This is the first mechanism that is applied where any kind of a squeeze is suspected. This usually occurs when there is an abnormal backwardation focussed on a short period. It takes experience as well as oversight of members' books to tell when an apparently impending squeeze is no more than a storm in a teacup, and when it is something potentially more sinister.

It must also assure transparency of its prices. In the five minute rings allocated to each metal four times a day, the open outcry trading method is still considered the finest way of achieving transparency. However, as a global market, the LME maintains official trading among members' offices throughout the London business day and clients connected to the Reuter screen can see the prices posted by at least a dozen ring members. (Every ring dealer is, of course, a market maker). Closing prices from the LME's two ring dealing sessions a day are distributed worldwide by the Exchange's licensed quote vendors, including major wire services like Reuters, Knight Ridder and Bloomberg.

As mentioned under members' regulation, the LME is itself part of the paperchase assuring that a record is kept of every transaction that is made across the ring and cleared through the LCH. To assure its rules being obeyed at all times, and to resolve disputes between members (which often needs to be done quickly), the Exchange has its own Compliance Officer. Also, which is much more than symbolic, the Exchange premises have always been laid out so that the Boardroom is adjacent to the Ring. Thus if the intervention of a Board member is required at any time, this is immediately available; normally, a Board member is present on the Ring during dealing hours. Similarly, the Exchange is well placed to arbitrate disputes that may arise between ring dealers and their clients.

IF THE WORST HAPPENS

We have touched in Chapter 6 on the machinery by which a cleared exchange like the LME is protected against the potentially disastrous effects of a ring dealer defaulting due to going bankrupt or for any other reason. Provision of such machinery is not only commercially necessary, it is also required by regulation. This means the exact mechanism – who does what and when – for the closing of the open positions of a defaulting ring member and contingent actions must be clearly agreed in advance between the Exchange and its clearer. Such, indeed, is the position as it has been set up between the LME and the LCH. Contracts would not be allowed to run to maturity, nor would any physical metal be moved. Rather, as on all other commodity futures markets, settlement would be by way of payment of differences arising from marking the open positions to market.

In many ways, what is more to be feared by a ring dealer than the failure of one of his contemporaries is a failure of a client. This eventuality is not covered by the clearing house, and the regulator's view is that he has ensured the broker is financially strong enough to withstand such a blow. True, under the terms of his contract with his client, the broker will have similar rights to those of a clearing house to liquidate his defaulting client's open positions without waiting for any further instructions from the client. But if the client has been on the wrong side of the market for some time and is near the limit of the credit his broker extends to him to meet margin calls, the broker is certainly facing a loss. This, again, is why brokers appear to have rather sweeping powers to vary the credit they will extend to a client. If the alarm bells begin to jangle on the rumour mill, it is better for the client as well as the broker if the first action is the broker restricting the client's credit, rather than it be the client defaulting.

Whereas the powers of a clearing house to square the position of a defaulting ring dealer can, by law, override normal insolvency orders of priority, the broker has no such protection when dealing with a defaulting client except in the relatively rare instances where property has been put up as collateral. On this class of asset, the broker's claim ranks before others.

In the cut and thrust of inter-exchange rivalry, Nymex, as the new parent of Comex, attempted to claim that LME was less tightly regulated than markets in the USA, this despite a high degree nowadays of mutual recognition of the merit and enforceability of each other's regulation across the Atlantic. In response, the LME used the following chart in its advertising to show the extent and strength of the safety net underlying its operations:

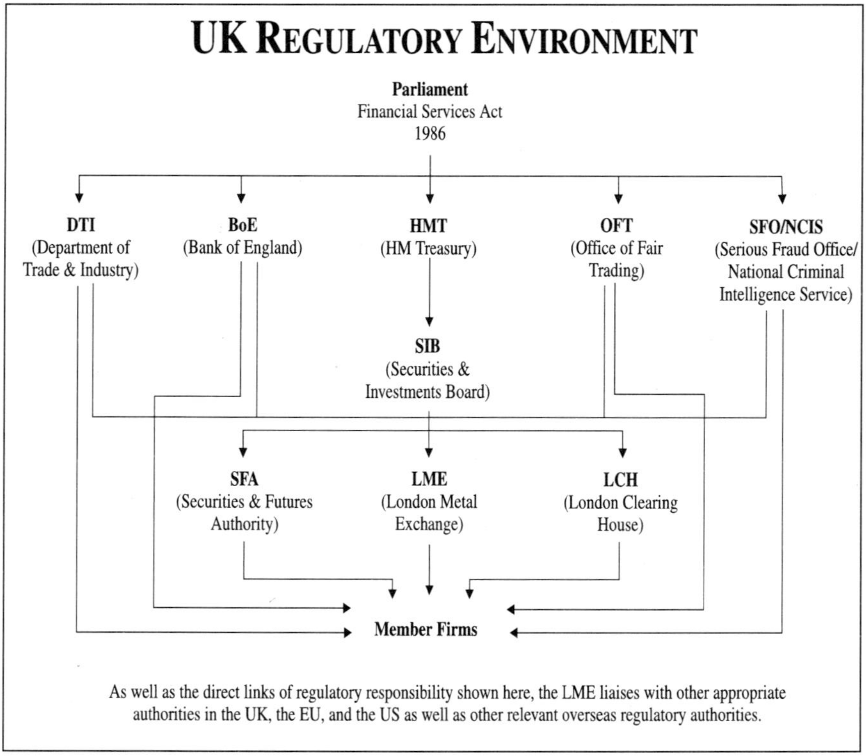

However, it is from Europe rather than the USA that the next wave of regulatory pressure will impact futures trading in the UK. With effect from January 1 1996 two EU directives will be overlaid on the existing UK law. These are the Capital Adequacy Directive and the Investment Services Directive. Like the Financial Services Act, they are mainly directed to the protection of individual investors in retail markets, but inevitably fail sufficiently to distinguish professional operations such as the LME, its brokers and their clients. One advantage these Directives do have over the FSA (on which they are otherwise extensively modelled) is that they do not attempt to drag commodity or commodity futures trading into an arena which is predominantly concerned with individual investors and equity trading. However, only a minority of LME ring dealers are so limited in their brokerage services that they are dealing exclusively in commodity futures. Even if they have a minority activity accounting for only a few percent of their total activity which is not commodity trading within the meaning of the Directive, then they will be subject to these directives.

The Investment Services Directive is, as indicated, very similar to the FSA, but has some subtle differences due to taking in some points from the rulebook of the COB (the French equivalent of the SIB). If a firm is subject to the ISD (and, as indicated, most LME Ring dealers are, in one way or another, subject to it), then it is also caught by the Capital Adequacy Directive. Again, UK rules on this point (implemented, in the case of LME firms, by the SFA) are virtually the same as the CAD, which was modelled on them.

There are a number of good things about the new Directives beside their recognition of the difference between equities trading and commodity futures trading. One is that they equalise more effectively than at present the competitive situation between commodity brokers and banks when the latter are seeking to move into the former's territory. This is achieved, not through the Directives themselves, but through two parallel Directives, the Solvency Ratio Directive and the Large Exposure Directive which effectively bring commodity brokers in line with what is – and is not – imposed on banks through the second Banking Co-ordination Directive.

Another piece of playing field-levelling the Directives achieve is on offering brokerage services throughout Europe. At present a brokerage house which is a force throughout Europe has to have subsidiaries in each country of the EU where it wishes to do business; these then remain subject to the national jurisdiction of their parent's domicile country, so are effectively dealing offshore in the country in which they are based! Under the ISD firms will be able to establish branches in other EU countries and be treated, and regulated, as far as their operations are concerned, the same as local firms. This includes scope for the branches to become members of the national exchanges of their country of domicile. However, trading supervision, including ensuring compliance with capital adequacy rules, would come from the parent's home country. This concept enjoys the bureaucratic designation of superequivalence and is rather reminiscent of George Orwell's "Animal Farm" in which appeared the slogan "All animals are equal, but some are more equal than others".

CHAPTER 9

The physical side

Many things attest to the character of the London Metal Exchange as a trade-oriented futures market, and to its conscious policy of continuing to have this identity. None, however, is more important, or more obvious in comparison with other commodity futures markets, than the attention the LME gives to the metal that underlies its contracts and to the relative ease with which it can be moved into and out of its warehouses. All this, however, is not to deny the fundamental truism that, from the viewpoint of the physical metal trader, a futures market is his source and outlet of last resort.

As far as concerns the quality of the metal that may be delivered against LME contracts, the specifications are contained in Appendix II. Additional information on particularities of these contract qualities is given, where necessary, in the chapters on the individual metals, while Chapter 19 on the history of the Exchange gives insight into the long and at times tortuous road that has brought the story thus far. It also teaches us never to assume that any one element of LME trading will be set in stone for ever.

In an age of technology and standardisation, the LME's relatively high dependence on individual brands may seem at times a little old-fashioned. This is not to say that reference is not made to relevant international specifications; it frequently is. In the case of a contract like aluminium alloy, it is fair to say that the contract could never have been introduced without a framework of established internationally-accepted standards in place first.

However, specifying brands has two advantages. In the first place it harnesses the commercial commitment of the individual producer to ensuring that its metal meets its regular standards. It would be foolish for a producer to think that it could produce a batch of metal to the barest minimum of the relevant standard because it was only going to be delivered to an LME warehouse. Sooner or later the metal would be delivered out for industrial consumption and at that time its worse-than-usual analysis would immediately be noticed. It would then be an embarrassment for the producer of the brand, not for the LME or any of the possibly many dozens of intermediate owners of the warrant. To underpin this commercial pressure, the LME's criteria for the acceptance of brands for registration include a written commitment from the producer that its production will consistently match up to the quality of the samples provided for testing at the time of applying for registration.

The other advantage of featuring brands as a defining criterion is that this in fact allows LME-deliverable metal to keep abreast of subtle changes in industrial practice which national specifications may not have the flexibility to reflect as soon as they become established.

The normal procedure for making a brand deliverable on the LME is by the producer applying for registration, as explained more fully in Appendix III. However, when a new contract is being introduced there may not be sufficient producer enthusiasm for the concept to assure a broad base of registered brands at the outset. In this case, the Exchange will simply list the brands that are deliverable, limiting itself to the best-known and most reliable. When this happens, the procedures for LME assayer analysis of samples and the industrial testing of samples by two well-known consumers are not applied. Once again, this is an area in constant flux; some LME brands have been registered from time immemorial, but there is also a continuous process of adding new brands and deleting old ones.

WARRANT SWAPPING

With sellers having the option to select from their holdings the brand and location of the warrant they will deliver, it is quite common for a buyer who wants to take up his purchase as physical to find that what he is getting is far from suitable. This is why there is a flourishing business among brokers in warrant swapping to secure for their client a brand and location that meets his needs. Like any other service, this has to be paid for by way of premium, and these can at times be quite considerable, up to the full cost of shipping an unpopular warrant from a remote warehouse and re-refining it to top specification or blending it into other parcels of premium metal. All this leads to a sub-culture of knowledge of the best and worst brands and locations, and most market operators review their holdings of warrants regularly to try to hold all the winners and none of the losers.

SINGAPORE

C. STEINWEG

PORT OF STORAGE : SINGAPORE
LOCATION: PT 10

L.M.E. Warrant No.: SG832682 **IN BOND**

IS ENTITLED TO RECEIVE:

BEARER

26(twenty six) bundles = 1144(one thousand one hundred forty four)
SPECIAL HIGH GRADE ZINC INGOTS

BRAND : HX

weighing 24788 Kilograms Nett

Not insured by us, insurance is the responsibility of the bearer of this warrant.

Storage rent per 1000 kos. per day: USD 0.15 commencing: 03/~~Jun/1994~~

payable on delivery.

Form.No.: 0315821 Dated London: 03/Jun/1994

C.STEINWEG
WAREHOUSING (F.E.) PTE LTD
37 JURONG PORT ROAD
SINGAPORE

Agents in London:
C. SHAW LOVELL & SONS Ltd.
LONDON
Tel.: (01) 6265711
Fax: (01) 9291451

For conditions p.t.o.

We have already mentioned the importance of warrants in LME trading as the key linking piece of paper between the futures trade and the underlying metal. They are also a bearer document, so need suitable security in handling, plus they must be delivered in the clearing to a deadline. Saying "The warrant is in the post" is not acceptable! Nevertheless, like any piece of paper, warrants do occasionally go astray and a duplicate can be obtained from the issuing warehouse at the cost of some trouble and expense.

IMPORTANT WAREHOUSES

It reflects the importance of the management of the physical metal that underlies LME trading that the Exchange relies so heavily on its registered warehouses for the services they supply. Because of the need for these services to be delivered 100% reliably, a good deal of Board activity and even more secretariat time is devoted to warehouse matters. What was as recently as the fifties a UK-based service involving some two dozen companies is now, as the listing in Appendix V shows, a worldwide business with a cast of hundreds. Becoming an LME-registered warehouse involves acceptance procedures (including physical inspection of the premises, weighing, record-keeping and materials handling facilities) that are no less stringent than those applying to brands.

In the chain of movement of metal through warehouses, a warehouse's first responsibility is to control the metal delivered for warranting. As with any transaction, this involves weighing, checking the integrity of bundles and checking brands and other producer markings. In the case of nickel, the warehouse is normally obliged to empty each drum, inspect and weigh the content and re-drum and re-seal unless the metal has been warranted by a Ring dealer. In this case only a 10% sample need be subjected to this process.

If, as is quite likely, the metal is not already in warrant-sized parcels, it is up to the warehouse to make them up. The warehouse is then in a position to issue a warrant and to start storing the metal under the conditions required by the LME. Although these conditions may be quite onerous (as in the case of aluminium alloy, for example), the standardisation of LME processes means that warehouse rents are quite reasonable compared with an individual transaction for the same service. Naturally, charges are made for accepting metal into warehouse (paid by the supplier) and delivering it out (paid by the warrant-holder). But it is the warrant that doesn't move and quietly clocks up rent that is most attractive to a warehouse owner.

Warehouse record-keeping is critical on two counts. The first is, obviously, that when a warrant-holder elects to take up the metal, the metal must be there in full measure (plus or minus the 2% allowance in the LME contract) and complying with the pieceweights, brand and any other identification listed on the warrant. The other aspect is the twice-weekly reporting to the LME on total stocks held, movements in and movements out. Again, because of the trade-oriented nature of the LME, the total statistics of warehouse stocks, and the movements in individual warehouse locations are vital market information. They are studied just as intently as other fundamental statistics like production, consumption and industry stocks. And unlike those statistics, LME warehouse stocks figures are almost instant.

Warehouse rents are the subject of some unusual but neatly effective conventions. They are charged, in US cents per ton of metal, on a weekly basis at rates agreed with the LME for the particular warehouse company and location, so rents are not identical for all warehouse locations. Warehouses must give the LME three months notice of a change in rental rates. For warrants that do not move, rents accrue from week to week and are settled once a year in March. Obviously, during this period the warrant may have been owned by many people and this is taken care of by a neat arrangement. Each time the warrant changes hands, the seller allows the buyer the rent already incurred up to the moment of sale. Thus by a process similar to VAT inputs and charges, each owner of the warrant pays rent only on the period he has owned it. As indicated, the accrued rent is cash-settled with the warehouse in March by whoever is holding the warrant at the time. If the warrant is taken up as physical, accrued rent becomes payable at that moment.

Because of their more or less cast iron security, warrants are accepted as collateral, not only in relation to LME trading, but also for any credit or capital adequacy control purposes and indeed for most banking purposes. A point to bear in mind where timeliness after putting metal on warrant is concerned is that there is a difference between the older metals copper, lead, tin and zinc on the one hand and the newer metals nickel, aluminium and aluminium alloy on the other. For the first group, the warrant is issued directly to the person putting the metal on warrant. For the latter group, however, where a clearing broker is required to have some responsibililty for the integrity of the metal delivered to the warehouse for warranting, the warrant is issued to the order of that clearer and only reaches the person putting the metal on warrant when it has been endorsed in his favour by the clearer.

A reflection of the important place occupied by warehouses in linking the LME to its physical metal hinterland was provided in 1995 when the Exchange opened new warehouses for copper in the USA. Previously, its only US warehouse facilities had been for the other five LME primary metals. One reason for skirting round the question of physical copper stocks in the USA had been the dutiable status of the metal in that country, though this also applied to lead and zinc. However, another and more potent reason was the politically-sensitive question of the existence of the US-based Comex market, with its own warehouse network. There is no doubt that as the industry discovers over time the advantage of having LME warehouses near to hand, the use of those warehouses grows and with it the use of LME trading.

ASSAYERS TOO

Perhaps nothing illustrates the overlap between the LME and physical trading more than its official recognition of the work of samplers and assayers. Officially-recognised assayers are continuously called on to check samples of metal moving into or out of LME warehouse, as well as to report on samples of brands applying for registration. With the zinc contract, for example, calling for minimum 99.995% purity, it can be understood that a high degree of professionalism is needed for an assay at this level of accuracy. If a warrant is ever rejected after being drawn down from warehouse, it is almost certain that an LME assayer will find himself in the hot seat. He could perhaps also share it with one of the international surveillance companies if the basis of rejection had more to do with the physical condition of the metal than its assay. Fortunately, such instances are extremely rare.

HOW MUCH DOES 25 TONNES WEIGH?

We have referred earlier to the scope for banking operations on the LME when a metal is running a contango between any two positions which represents a better return on money invested in that metal than is available in the money markets. We referred to such transactions as watertight, because the nearby purchase and forward sale are executed at the same moment, locking in the contango and making the deal immune to subsequent fluctuations in the market. All this is true, provided only that the bank and the broker servicing it remember the potential catch in such a transaction from the 2% up or down tolerance on weight allowed by LME contracts and the way any variation within this tolerance is accounted.

For the sake of simplicity, we will suppose a bank instructs its LME broker to buy a warrant of copper for cash and sell it for three months as the prevailing contango represents a 2.5% gain, or a 10% annual return after deducting warehouse rent, insurance and broker's commission. These items might total 1.5% of the contract value over the three months, or 6% annualised. The broker buys a warrant in the market at $3000 for cash and sells it for $3120 three months – a 4% gain. On the face of it the bank has achieved a clear 10% annualised return against a best money market rate of 9%. Good business. However, the warrant turns out to be 2% underweight – just within tolerance. This is where it is important to know that when an underweight warrant is delivered, the short weight is priced at the then prevailing settlement price. If the market were to rise over the three months, so that by the time the forward sale matured settlement was $3500, the exact calculation would then look as follows:

Buy 1 warrant basis $3000	$75,000
Less shortweight basis $3000	$1,500
Cost	$73,500
Sell 1 warrant basis $3120	$78,000
Less shortweight basis $3500	$1,750
Realisation	$76,250
Less brokerage etc	$1,102
Profit	$ 1,648

Equals 2.24% or 8.97% annualised, not quite so wonderful. Likewise, an overweight warrant would be bad news in a falling market. Another variable that needs to be borne in mind is warehouse rents – rates at some warehouses can be almost double those at others. In practice, the majority of transactions of this type are entered into for periods shorter than three months, sometimes as little as two or three days.

This problem can be avoided if the client tells the broker what his effective interest rate objective is and leaves it to him to produce a guaranteed yield. It is up to the broker to use his professional skills to avoid falling into the overweight/underweight trap. If on the other hand the deal is done as an execution for the client, there is still one way in which he could cause problems for his broker. Because it is central to this type of operation that the same warrant be bought and sold, and because clients may lose control of an individual warrant in a portfolio situation, or may use it as collateral, brokers protect themselves against the embarrassing possibility of trying to sell a different warrant from that which they bought by getting the client's agreement to a rider clause on the following lines:

"Metal covered by this contract must not be disposed of except by relending to the market through Rudolf Wolff and Co Ltd. Thus if you wish to terminate your position earlier than the forward sales date in order to regain your funds you will, through us on the LME, repurchase your forward date and sell the cash (spot) metal. Against the latter you deliver us your warrants thereby bringing forward the date of your final settlement".

ARBITRAGE

Arbitrage is another type of operation that crosses the divide between physical and futures, and is again one where a full knowledge of the physical side of the equation is vital to a successful transaction. The classic definition of an arbitrage deal is the simultaneous purchase of a metal in one futures market and its sale in another to profit from the relationship between the two markets after covering the costs of brokerage. For all practical purposes in a metals context, we are talking here purely of the LME, Comex and copper as there is no other futures market in any metal with anything like the liquidity of the LME. Without a clear knowledge of the cost factors specific to physical copper trade in the USA and outside it, it is easy to misread the true scope of an arbitrage.

Indeed, it helps set the scene for a futures arbitrage if we first describe the sort of physical arbitrage questions that face producers, merchants and some of the most enterprising consumers every day. Let us envisage a Chilean producer who has uncommitted tonnage of copper he can ship to either Europe, Japan or the USA. His costs and revenue picture is different in each market, due to freight differentials, tariffs (or lack of them), physical premia over the relevant futures reference prices and of course the reference prices themselves, the USA traditionally referring to Comex and the rest of the world to LME. Taking a base line of shipment to Europe and a sale basis LME prices, all the alternatives can be expressed as differentials.

Thus compared with freight to Rotterdam, it costs him 0.25 cents per lb more to ship to the USA and 0.75 cents more to Japan. His physical premium in Europe is, say, 0.2 cents and in Japan 1.15 cents, both these over LME. In the USA his premium is 2 cents over Comex, which we will assume is 2 cents below the LME. Last but not least, the USA charges a duty of 1 cent per lb. He then sets out his options as follows:

	Europe	**Japan**	**USA**
Costs			
Freight	0	(0.25)	(0.75)
Duty	0	0	(1.00)
Revenues			
Premium	0.20	1.15	2.00
Arbitrage	2.00	2.00	0
Net	**2.18**	**2.90**	**0.25**

No prizes for guessing where he will send his metal that day! However, next month, or whenever he has disposable tonnage, he will have to set out his options all over again. Apart from the movements in the futures market reference prices that will undoubtedly have taken place, there may well have been subtle movements in physical premia in one, two or three markets. Even duties are not a constant in these Uruguay Round days. Strictly speaking, Japan has an import duty which rides with fixed levels of price, but it is so long since prices were in those fixed ranges, most people have forgotten about it.

USING THE ARBITRAGE AS A HEDGE

This leads us to consider the physical copper merchant whose business is moving parcels of material around the world. It is normal for him to buy in the pricing sphere of one market – say Comex – and sell in the pricing sphere of another, the LME. If he is also entering a commitment for a forward transaction with pricing on the respective markets at a future date, he could be in double jeopardy if both markets were to move against him in the interval between entering the commitment and implementing pricing. This is the more so as his buying and selling dates are separated by the time it takes to ship the metal. He therefore locks in his differential between the two markets for the respective dates by an arbitrage.

Historically, it has been normal to find the LME standing at a 2-3 cents premium to Comex, which would favour a transaction buying basis Comex and selling basis LME. Whether this will remain a norm in the future has yet to be proved. The opening of LME copper warehouses in the USA may have the effect of keeping the markets closer to each other, or in other words narrowing the arbitrage.

For the purposes of this illustration we will assume our merchant is buying cathodes fob west coast USA (and therefore basis Comex) and selling them in Japan (basis LME). There will also be physical premia to consider – they may well be the basis of the profitability of the deal. He also, of course, has to factor his shipping, insurance and financing costs into the equation. To keep this example simple we will assume the metal is being bought and sold at physical premia which pay these costs, so all these items can be ignored. We will now assume it is June and September Comex is trading at 150 cents per lb with October LME dates (Wednesdays) at 152 cents. These months correspond to his planned shipping and delivery dates, so he needs to lock in this favourable differential by an arbitrage. He therefore buys September Comex and sells October LME.

The transaction can now be viewed as, in effect, two parallel futures hedge operations. Let us assume, to really underline the point, that between June, when he enters the commitment to the physical trade, and September when he prices his purchase, Comex moves up 2 cents, and between June and October, when he prices his sale, LME falls 2 cents. It is clear the physical transaction pricing now produces a 2 cents per lb loss instead of the 2 cents per lb gain prevailing between these forward dates in June. Thanks to the arbitrage, however, that favourable relationship has been locked in and contributes to the profit on the deal.

As always, the broad outlines do not tell the whole story. If we have effectively set up two hedge transactions, there are two sets of brokerage to pay and, certainly as far as Comex is concerned, potential margin calls to be considered. Equally importantly, the deal needs to be structured to cater for the different lot sizes on LME and Comex. Whereas the LME lot is for 25 metric tonnes +/- 2%, a Comex lot is 12.5 short tons. The simple answer to this is to say that five LME lots equal 11 Comex lots. Today's reader of this book who thinks this is complicated should reflect that in earlier editions it was necessary also to explain how to cover the additional exposure due to the two markets operating in different currencies and on quite meaningfully different quality specifications! Fortunately these elements are not problems today.

CHAPTER 10

Pricing systems for base metals

Many individuals go through a lifetime in the metals producing and consuming industry without once entering into a London Metal Exchange transaction. Yet the LME prices are absolutely essential to their commercial activity. The connection between the nerve centre on the LME and these substantial limbs of the industry is provided by pricing.

In an industrial context, the main commercial considerations for a metals buyer are:

1) a reliable and timely supply of metal of correct quality for the works' processes.
2) a price basis for those supplies which ensures that he is never at a disadvantage with his competitors.

Both these criteria are met by suppliers and consumers entering into long term contracts for the supply of metal in which the value of each shipment is indexed on the appropriate LME price. To an economist, the LME, its ring-dealing members and their clients are, as a vital by-product of their dealings on the exchange, engaging in price discovery. In other words they are price-making.

Nobody knows a few minutes before the close of the LME second ring exactly what the value of that metal is at that moment. A few moments later, when the ring-dealing process has finished for the time being, a closing price is produced and is accepted worldwide as the most valid indicator of the value of that metal at that moment. The parties involved in setting that closing price are, therefore, engaging in a price-making activity, even though it may not be the purpose that is uppermost in their minds at the time of entering the transactions.

At the same time, reference should also be made to those ring-dealers who may be vociferously bidding to buy or sell 'on the bell' with one eye very firmly fixed on the impact they are having on the price that will eventually be quoted. They are still engaging in a process of price-making but, if somebody takes up their bid or offer, then they have entered into a transaction and this may not be what they primarily had in mind, though it will certainly reinforce the price-making nature of their trading.

It follows that, if the LME is where the price-making is done, it is in the sales and purchasing offices all around the world where period pricing contracts are being managed and implemented, that price-taking takes place.

A typical annual contract for the physical supply of metal will specify the tonnage to be delivered each month and the formula by which the value of that material is calculated using the agreed relevant LME price as a variable.

Behind this simple statement lies a mountain of intensely complicated practice, so that the negotiation and management of a period pricing contract has become a distinct commercial skill. When the LME may be liable to rise or fall by, for the sake of example, \$100 in a month, it may be wondered why grown men and women spend hours in smoke-filled rooms haggling over the last dollar on a differential over or under the LME reference price in a period contract.

The answer, of course, lies in the scale of such contracts. It is nothing for a large consumer to contract for one, two or three thousand tonnes of, say, copper a month. At this level, even a \$1 a tonne variation on the margin of difference from the LME price amounts to a goodly number of dollars at the end of the year. We should also respect the psychological angle that, with most of the variability of price being taken care of by the ups and downs of the LME, the fixed differentials written into the contract become the focus for negotiation!

DIRECT DELIVERY V LME WAREHOUSE SUPPLY

What are these differentials and how do they arise? Essentially a period-pricing contract attempts in its structure to cater for the fact that a direct producer/consumer transaction in physical metal has some noticeable differences from the situation that would arise if either the producer sold his metal on the LME or the consumer bought it there.

For the producer, a sale on the LME involves shipping metal to the nearest LME warehouse location and paying the cost of placing the metal on warrant. If the producer's brand has a particular characteristic that is highly favoured by some consumers, such that he is able to persuade them in the physical market to pay for that quality, then he would forego this by selling on a standardised market like the LME.

Conversely, the consumer would have to pay the costs of taking the metal off warrant and transporting to his works. He would also have to accept as good delivery, at seller's option, any LME brand lying at any LME warehouse. This might include brands which experience has told him are definitely less suited to some of the more technically sensitive areas of his manufacturing process. Likewise, it takes time and costs money for his broker to undertake the necessary warrant-swapping to secure metal for him in a warehouse reasonably close to his works. So it makes perfect sense for the producer and consumer to get together and agree to ship physical metal from the one directly to the other.

It also makes sense for them to agree to price the metal on the basis of the LME because both parties then know that they are giving nothing away in terms of basic reference price to their competitors.

In the case of copper, the formalisation of pricing processes has proceeded to the point where both the reference price and the main standard premium for physical delivery are regularly quoted in Metal Bulletin. Period pricing contracts with the main African copper producers specify that the metal will be supplied on the basis of the CIF Europe producer price prevailing on the day or during the period in question. To ask the consumer to buy copper on the basis of a price nominated by the producer might represent something of a pig in a poke, were it not also understood that the producer price on any day is that day's LME official settlement price. However, it has not always been so. For a number of years the producer price was the same as the three-month sellers'.

Having defined the moving copper price that forms the basis of the invoice calculation, the copper period pricing contracts also specify the size of premium for direct delivery of producer copper which in itself varies according to the pricing terms available to the consumer. For a detailed explanation of all the ramifications of a period pricing contract, readers are referred to 'Trading in Metals' also published by Metal Bulletin Books, but the following summary suffices here.

Nearly all period pricing contracts are written for one year. In the case of 'evergreen' contracts, all the basic terms remain unchanged from year to year unless altered by the principals in negotiation.

THE MATING SEASON

The practice of writing period pricing contracts for a calendar year is so entrenched in many sectors of the metal business that the period from about October to December each year is known as 'the mating season' when major producers and consumers negotiate furiously over the small print of the contracts that will govern shipments of metal over the ensuing year.

Within the year, the most common structure of period pricing contract calls for regular monthly deliveries of a given quantity.

The producer may have some options to vary the tonnage within any one month provided he catches up or averages down to the agreed rate of delivery within one or two months thereafter. If a strike is threatened because a labour contract is coming to an end, then producer and consumer may find it in their mutual interest to ship as much metal as possible before the deadline and disregard the fine print of the contract!

The most elementary pricing formula to be found in period pricing contracts says that the month's tonnage will be priced at the average of the LME price of the month of shipment. Scarcely more complicated are the contracts which use the average of the month prior to the month of shipment or month following month of shipment. But 'average' is not a simple term. It may be the average of the nominated producer price as we saw in the instance of copper above. It may be the average of the mean of three months, or the average of cash buyers or the average of cash sellers, etc. etc. It may even be the average of 'the mean of four' which means that every day some poor clerk – or more probably a software programme – has got to add up cash buyers, cash sellers, three-months buyers and three-month sellers and then divide by four to arrive at a single figure.

More complicated contracts, found particularly in copper, accept that this is a high value volatile commodity where the date of delivery is of importance and the rate of delivery is, therefore, set in terms of so much per week or even per day. In this last case, the consumer 'draws down' his permitted tonnage from the producer's stocks on a daily basis and they are priced accordingly. Actual delivery may not necessarily take place the same day; it is sufficient for producer and consumer to know the extent to which their book has been adjusted following that day's transactions.

ANOTHER COMPLICATION

We referred above to African copper producers operating two levels of premium over their LME-related reference price. One is pricing on the average, which is as described above, and is associated with the lower of the two premia in the structure. The other is pricing on the unknown.

Unknown pricing means pricing on the basis of the same day as the copper is drawn down ie the unknown LME official price of an hour or two later as distinct from the known price of some 22 hours earlier. On top of this, the contract may also provide for the consumer to vary the level of his daily intake within certain tonnage limits; this is known as the 'swing'. Whilst the roots of this facility lie in the unpredictability of the volume of offtake the consumer will experience from his customers during the day, it has become a pricing facility capable of advantageous manipulation by a skilled and adventurous copper buyer. Elements of these approaches to operating a period pricing contract may also be found in other base metals; for real complexity, nothing can match a zinc concentrates contract.

Many of the parties to the more sophisticated period pricing contracts are themselves also sophisticated users of the Exchange for hedging and price control. However, it must never be forgotten that hundreds of transactions take place daily, priced on the basis of LME price, without ever once producing any business for the LME, either in the shape of hedging, or of physical movement of metal through LME warehouses.

Conventional wisdom of the scrap trade, for example, is that nearly all grades of scrap are in practice unhedgeable on the LME because of the differentials between scrap prices and virgin metal prices, which may change quite dramatically in a short period of time and confound the best laid hedging plans.

Nevertheless tens of thousands of tonnes of scrap are sold and priced every day under factory clearance and other period pricing contracts on the basis of LME prices.

A small industry is at work supplying this price information to the metal trade all the way through from the speed of a wire service like Reuter, Knight Ridder or Bloomberg, to sending the boy down to the street corner newsagents to get a copy of yesterday's financial paper to find the day before's LME price. However, it should not be thought that this relationship between these more remote price takers and the LME cannot be to their mutual advantage. By their unquestioning acceptance of the authority and validity of LME prices as the standard yardstick of value of LME metals, these price takers are performing a valuable function in reinforcing and underlining that authority of LME prices. Without such authority, the backwash of hedging and other transactions into the LME ring every day would be much less.

CHAPTER 11

Reading the market

Formal markets and analysis go hand in hand. When Adam Smith postulated the law of supply and demand, he had to assume that the scale and trend of supply and demand were known factors. To this day, fundamental analysis, as it is called, relies principally on an assessment, both statistically and in other ways, of supply and demand. Get that equation right, fundamental analysts say, and you will get the right idea of the "right" price for the metal in question. Certainly, you should get an idea of the trend, up or down.

The whole point of having an idea of the " right" price, of course, is to spot when the market price is "wrong". A miner may be making five to ten year projections of likely price levels as part of a complicated discounted cash flow projection for a property the company has already spent millions on proving; it now has to decide whether, and if so when, to start developing it for production. To him, the key figures are likely annual average levels of price for the metal in question. He is quite unconcerned with the day to day oscillations of prices that preoccupy all brokers and many of their clients. But if he gets his long term readings of fundamentals and prices wrong, and the price is not going to be "right" for the economics of the project when it comes on stream, it could ruin the company. One of the easiest ways to get a long-term forecast wrong is to assume that today's price is "right", when in fact it is "wrong" (and vice versa).

Notice that we are referring here to supply and demand, not to production and consumption. These statistics are, naturally, a key part of supply and demand projections. But even over as long a period as five or ten years, there is a further statistic which is critical to reading the supply-demand picture, and that is stocks. It can easily happen in investment metals like gold and silver that there is a secular trend in market stocks over a period of many years which will sustain (or, if you are not asking the right questions, mask) a production-consumption imbalance. Such a trend may not normally be as strong or as long in base metals, but it commonly exists, and is a key factor in turning production and consumption numbers into supply-demand projections.

Equally important is to realise that stocks figures are, according to the mood of the market, counted either way. In other words, if the market is rising, stocks tend to be tightly held by producers or merchants, thus forcing consumers to bid yet higher for available metal. In a bear market, on the other hand, consumers live off their stocks and buy at less than their rate of consumption, while producers keep pushing prices down to find a level at which they can turn unsold stocks into cash. In the six LME primary metals, to a greater or less extent depending on the metal, it is also vital always to remember that there is a scrap (recycled) side to the industry statistical balance. As the historical chapter underlines, very expensive mistakes can be made by assuming that the scrap side of any market is always self-balancing and therefore can be ignored in supply-demand analyses.

As well as essentially statistical data, fundamental analysis also pays much attention to news. An impending strike at an important mine or smelter, for example, is a visible impact on the supply of that metal. Markets will be sensitive to the ebb and flow of news about the prospects for such a strike first developing and then lasting. Similar news-based fundamentals would include war, riot, weather, natural disasters, political developments (especially if affecting trade or tariffs, but also, nowadays, environmental legislation), labour unrest at both producing and consuming works (whether first- or second-line consumers) and macroeconomic statistics from major economies.

TWO SORTS OF ANALYSIS

All this was fine for the more leisured age that lasted from the inception of the LME to as late as the 1950's. However, more recent years have seen an explosive growth in the practice and use of what is called technical analysis. Technical analysis says that it is an unnecessarily roundabout route to attempt to read prices via the fundamentals. If you believe, say technical analysts, that the market is efficient and that the price it produces clears the market, then you need look no further than the record of prices to read the future of prices.

There are two merits in this argument. One is that technical analysis rolls forward with the last available market price. There is no need to wait half a week for LME warehouse stocks movements; three weeks for IPAI aluminium production statistics; six or more weeks for WBMS base metal production, consumption and trade statistics; three months for key macroeconomic statistics from major developed economies; or six months for an intergovernmental study group report. The other merit is that, as we have seen elsewhere in this book, supply and demand are not the only determinants of price behaviour. There is also market sentiment – in other words expectation. This factor, like the fundamental ones already discussed, is also cleared through the market's system every trading minute of every day.

So the claim for technical analysis is that it is quicker and more comprehensive than fundamental analysis. However, after the 1994/5 incursion into the LME by major hedge and other funds, almost entirely on the basis of a reading of the fundamentals, it would be a brave man who would say that technical analysis is universally better than fundamental analysis. Both have their supporters; the foregoing overview gives some idea of a "horses for courses" breakdown of what to use when; and some people manage to keep both kinds of analysis in mind at once.

THE TECHNICALITIES OF TECHNICAL ANALYSIS

Fundamental analysis, as can be seen, can be understood and used by those with some knowledge of each metal market (such as is imparted in the later chapters of this book) and a healthy dose of common sense. The big area for debate in fundamental analysis revolves around the basic fact that speed and accuracy in the production of news and statistics tend to be mutually incompatible. In other words, the faster you get the news or the figures out, the more chance there is that they will be found later to be flawed. The same can be said of speed and comprehensiveness. The IPAI can get its aluminium production figures out consistently fast because it aims to produce very basic, regionally-aggregated production figures. WBMS produces more detailed national production, consumption and trade figures, and these naturally take longer to collate.

So it is certain that the market user will be offered – often for money! – a good deal of technical analysis; in today's market, a majority of market users will find that they need to understand this material. At the extreme, of course, are the managed funds which have set up computerised trading systems designed exclusively on the basis of technical analysis. They then buy and sell directly from their computer programme in reaction to the movements in prices.

All brokers' reports make mention sooner or later of the market approaching a "chart point". This means that those who study charts of market prices (known as chartists) are all seeing the emergence of one of the chart formations that are central to this form of analysis. So if, for example, copper is about to drop to $1600, and the charts say that this will make it clear that the market is in a particular bear phase, it can be expected that many computer-generated sell orders will hit the market as soon as copper touches $1600. Thus even if a market user is, himself, not a chartist, he needs to know what the chartists are doing to the market in the short term.

The quickest way to illustrate the general point is with an example. One of the most common chart formations studied by chartists relies upon plotting prices on two different timescales. One of these timescales might be daily price movements and the other the rolling one-year average (ie every day, the latest day's price is added to the list of prices, the oldest year-ago price is dropped and a new average for the past year is calculated). The shorter-term and longer-term parameters can, of course, be extensively varied according to the inherent characteristics of the particular metal, the prevailing economic circumstances and the feature the chartist is trying to detect. For some purposes the 10-, 20-, 30-, or 40-day moving average will serve as the short-term chart (against, say, a year's moving average). For others, one or other of these averages will be the long term chart against the daily movement. What the chartist will be looking for is a crossover between the graphic representations of the two series.

If, for example, the market has been depressed and is now rising in the short to medium term, one can expect the daily price plot to be below the rolling annual average plot, but to be converging on it. On a particular day, which may come sooner or later than the chartist was expecting, the two lines will cross. To the ordinary citizen, this might well suggest that it is now time to be cautious, as the rolling annual average may well be close to the fundamentalist's "right" price. But more likely, the chartist, based on his experience of studying many formations in a variety of metals and commodities, will announce that the market is now entering a bull run, and will treat the crossover as a buy signal. If enough chartists are plotting the same two series, quite a few of them will recognise the crossover simultaneously and their combined "buy" signals can produce, short term, something of a self-fulfilling prophecy. This in turn stokes up the signals coming back to them from the market. All this explains why the cynical argue that what is really needed for market analysis is a PhD in mob psychology!

THE PSYCHOLOGICAL ASPECT

In fact, this is quite pertinent. Certain price levels do impress themselves on the collective memory of the operators in, and observers of, a particular market. Often expressed as a round figure – \$1 per lb or \$1500 per tonne, for example – these levels can be seen from the charts as areas of turbulence in the evolution of prices. In this respect they are rather like a party of hikers on the moors. Normally, they proceed steadily along the path in single file. If they then reach a stream which has to be leaped over, they tend to bunch and mill around before some of them make the break and start encouraging the rest to follow. To a chartist it would be vitally important to know when more than half the hikers had leaped over the stream, as he would then be sure the party would continue travelling in its original direction. However, if an influential proportion of the hikers refused to jump, the probability is that the group on the far bank would jump back, and the group would set off in a different direction. To a chartist, it matters little which is the outcome; to him the important thing is to read the mob psychology right and convert that reading into a profit opportunity.

Having grasped the conceptual thinking of the chartist, the market user must then be prepared to learn at least some of his language. Expressions like "base formation", "break-out", "support", "resistance", "trend", "overbought", "oversold", "top" and "bottom" are fairly self-explanatory (but will be dealt with a little more fully below). Most people are, however, somewhat puzzled when they first encounter "head and shoulders", "double top", "momentum" and "oscillator", to name but a few. As with all trade terms, these are in fact designed to speed communication, not just to confuse the newcomer.

In the world of the chartist, prices that are not moving up or down decisively and, in the medium term, continuously are "moving sideways". As he plots the vertical zig zags over a period of a month or two, the analyst is looking for some kind of common denominator between all the short-term peaks and short-term troughs, which he will then describe as a "base formation". This means that he believes that the market is focussing on a range of prices constricted between a psychological floor level ("support") and a psychological ceiling ("resistance"). These support and resistance lines may turn out to be horizontal and parallel, or they may be inclined upwards or downwards, or they may be converging or diverging. Each of these possible combinations of pattern tells the chartist what he should probably expect next. A specific event, such as the breaching of a support or resistance line or a chart crossover, as described above, will give him an idea as to when the price will break out of its sideways phase into a decisive up or down trend.

Once again, the trade user of the market, looking for relative stability of prices and a quiet life, will not be happy to hear that a breakout is coming; but if it is, he had better know about it. The chartist, on the other hand, will be very happy – partly because it fulfils a prediction he has probably been making for some weeks and partly because it is the volatility of the market that keeps him in business! Once again, the psychological angle should be respected. The trade user of the market will probably see a base formation as a consensus between all the forces bearing on the market. Because of his abovementioned preference for stability, he will therefore see it as continuing for a long time. The chartist, for his reasons, will be wishfully thinking a breakout is due any day. Dialogue between these two views, provided it does not become unduly heated, can only do good to both of them. And if bets are laid on the eventual outcome in the shape of a lunch, at least both parties can eat it!

AN INEXACT SCIENCE

Post-hoc representations of classic chart formations are necessary to grasp the principles of reading charts, but can be very frustrating to the would-be chartist trying to make up his mind as he goes along. Is he, for example, witnessing the formation of a head and shoulders or merely an inclined resistance? A head and shoulders, it should be explained, is a graph of daily or other short-term prices which begins to rise, pauses on a resistance level, breaks out to a decisive peak, falls back to a support level roughly equal to the previous resistance level and finally fades away to the level at which the formation started. Clearly the further the formation progresses, the more obvious it becomes to everybody that they are in a head and shoulders situation.

It is those with the ability to call the left shoulder, and most importantly the top of the head, who stand to gain most from their chartism. By the time it gets to the right shoulder, the formation is hardly news. A head and shoulders can, of course, as readily occur in a bear market, but upside down. Other formations for which the chartist will be on the look-out include "flags", "triangles" and "double bottoms" and these are illustrated in the charts with this chapter.

Modern chartism tends to use several indicators at once to test the validity of the message that appears to be emerging from a reading of the charts. Apart from overlaying moving averages, a common technique is the use of point and figure charts. Familiarly known as noughts and crosses, these focus on changes of direction in trading and are designed to highlight periods of intense activity in the market. The normal time scale on the horizontal axis is replaced by a progression each time the market changes direction. Otherwise, market movements are reflected by crosses for each rise of a given size (eg $5) and noughts for each fall of the same size. With the chart progressing to the right only when the price changes direction, several days may be summarised in one column by a few noughts or crosses. At time of great uncertainty, however, there might be half a dozen reversals of direction in one day, taking six columns of the chart to represent.

Another technique is known as the RSI for short, but this does not stand for Repetitive Strain Injury. It stands for Relative Strength Index and is designed to show how far the market is overbought or oversold. RSI calculations usually focus on a period of 14 days, at a time when it is not clear what, if any, trend will emerge. It works by comparing the average of the up movement to the average of the down. If buying pressure is growing relative to selling, the index moves up, and down if the opposite prevails. It can also be used as an oscillator to detect when buying or selling pressure is losing momentum.

Stochastics is an impressive name for a technique which is also relevant for relatively short-term market reading. It works by comparing closing prices with the day's trading range and applies the observation that in an uptrend, closing prices are closer to the top of the trading range than in a flat market. Similarly in a downtrend, closing prices are closer than usual to the low end of the range. In the accompanying Chart 3, the day's high-low is represented by a vertical line and the close by a short horizontal dash. In the past, the LME's trading pattern of short rings in two daily sessions might have been thought less suitable for this type of analysis, but with official, cleared trading continuing all day in inter-office dealing, this distinction from, for example, US markets no longer applies.

Even more exotic market-reading techniques are beginning to come into commodity market analysis from the financial futures markets. These may rely on modern academic work in the field such as Elliot Wave theory or Gann theory. The one such economic theory you are unlikely to hear a technical analyst quote is the Kondratief Cycle. This says that the normal business cycle in the world economy, conventionally said to have a duration of four years, actually forms part of a longer cycle of very approximately 54 years' duration. In the up-side of the cycle, successive peaks of the normal 4-year cycle are always higher than the preceding peak. On the downside, the peaks are successively lower. At which point, one should recall JM Keynes' observation that "In the long run, we are all dead."

Chart 1

The head and shoulders formation identifies the meeting point of two opposite trends. A neckline is drawn between the two shoulders; a breach of the neckline implies a continuation of the new trend. A minimum target is then calculated by taking the distance from the top of the head to the neckline. Prices will then be expected to move the same distance down again at the end of the formation (see vertical dotted lines).

Chart 2

This chart of high, low and closing prices shows three chart formations – a Double Bottom, a Flag and a Triangle. The Double Bottom shows two days when the sell-offs have been reversed by aggressive buying. This indicates good underlying support where buyers outnumber sellers.

The bull Flag highlights a period of profit-taking after a strong move. A break through the top of the flag signals a continuation of the initial move.

The Triangle represents a period of trading where buying and selling are fairly evenly matched. However, the uncertainty of the market makes buyers and sellers increasingly nervous about missing the boat, so buyers buy at ever higher numbers, while sellers trade at progressively lower numbers. In this example, the break out of this range identifies the time when buyers outnumber sellers. Significant moves can be expected once a triangle has been broken.

Chart 3

The 10- and 30-day moving averages can be used to smooth day to day fluctuations in price. Crossovers in the averages can provide confirmation of a new trend and are therefore often used as buy and sell signals. This use of moving averages only proves profitable when the markets are trending.

Chart 4

Use of the RSI can add weight to chart analysis by providing other indicators to the likely price direction. In this example, the movement of the RSI above 80 signals that the market is becoming overbought. As soon as prices start to fall and this is confirmed by a downturn in the RSI, a fall in prices is signalled. Likewise, the bounce in the RSI from the oversold level confirms there is good underlying demand. Another use of the RSI is to compare its trend with that of the price. An early breach of the RSI trendline provides early warning of a loss in upward momentum in the price.

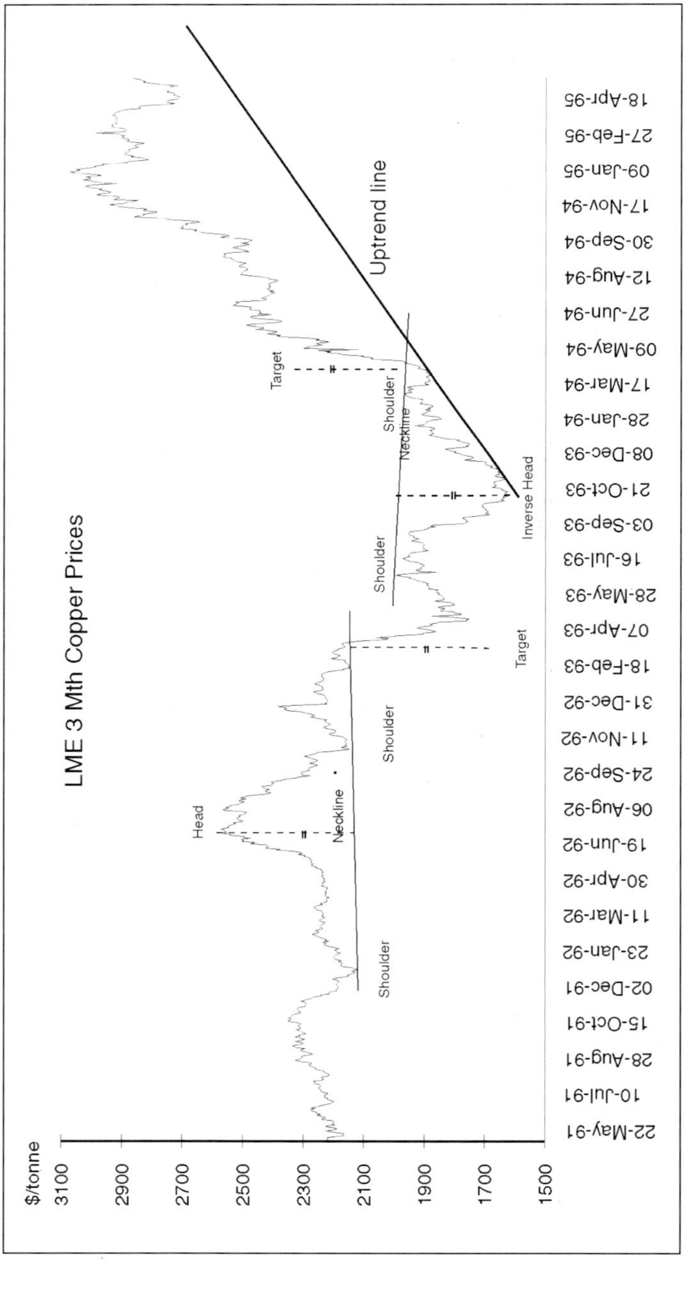
LME 3 Mth Copper Prices
$/tonne
3100
2900
2700
2500
2300
2100
1900
1700
1500
22-May-91
10-Jul-91
28-Aug-91
15-Oct-91
02-Dec-91
23-Jan-92
11-Mar-92
30-Apr-92
19-Jun-92
06-Aug-92
24-Sep-92
11-Nov-92
31-Dec-92
18-Feb-93
07-Apr-93
28-May-93
16-Jul-93
03-Sep-93
21-Oct-93
08-Dec-93
28-Jan-94
17-Mar-94
09-May-94
27-Jun-94
12-Aug-94
30-Sep-94
17-Nov-94
09-Jan-95
27-Feb-95
18-Apr-95
Head
Shoulder
Neckline
Shoulder
Target
Shoulder
Inverse Head
Neckline
Shoulder
Target
Uptrend line

Chart 1

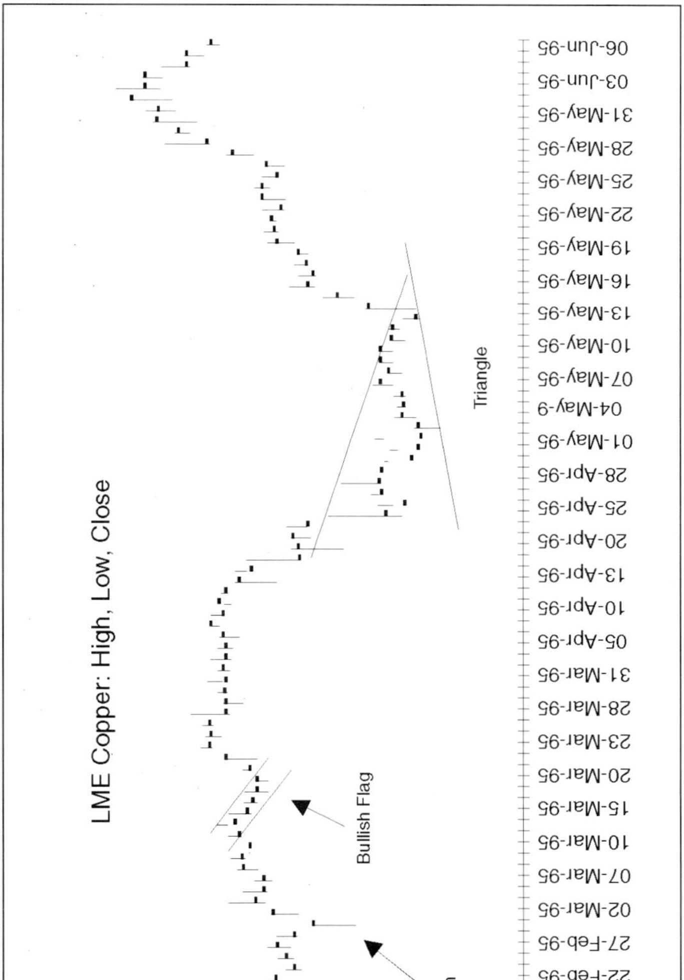
LME Copper: High, Low, Close
$/tonne
3050
3000
2950
2900
2850
2800
2750
2700
2650
2600
31-Jan-95
03-Feb-95
09-Feb-95
14-Feb-95
17-Feb-95
22-Feb-95
27-Feb-95
02-Mar-95
07-Mar-95
10-Mar-95
15-Mar-95
20-Mar-95
23-Mar-95
28-Mar-95
31-Mar-95
05-Apr-95
10-Apr-95
13-Apr-95
20-Apr-95
25-Apr-95
28-Apr-95
01-May-95
04-May-9
07-May-95
10-May-95
13-May-95
16-May-95
19-May-95
22-May-95
25-May-95
28-May-95
31-May-95
03-Jun-95
06-Jun-95
Double Bottom
Bullish Flag
Triangle

Chart 2

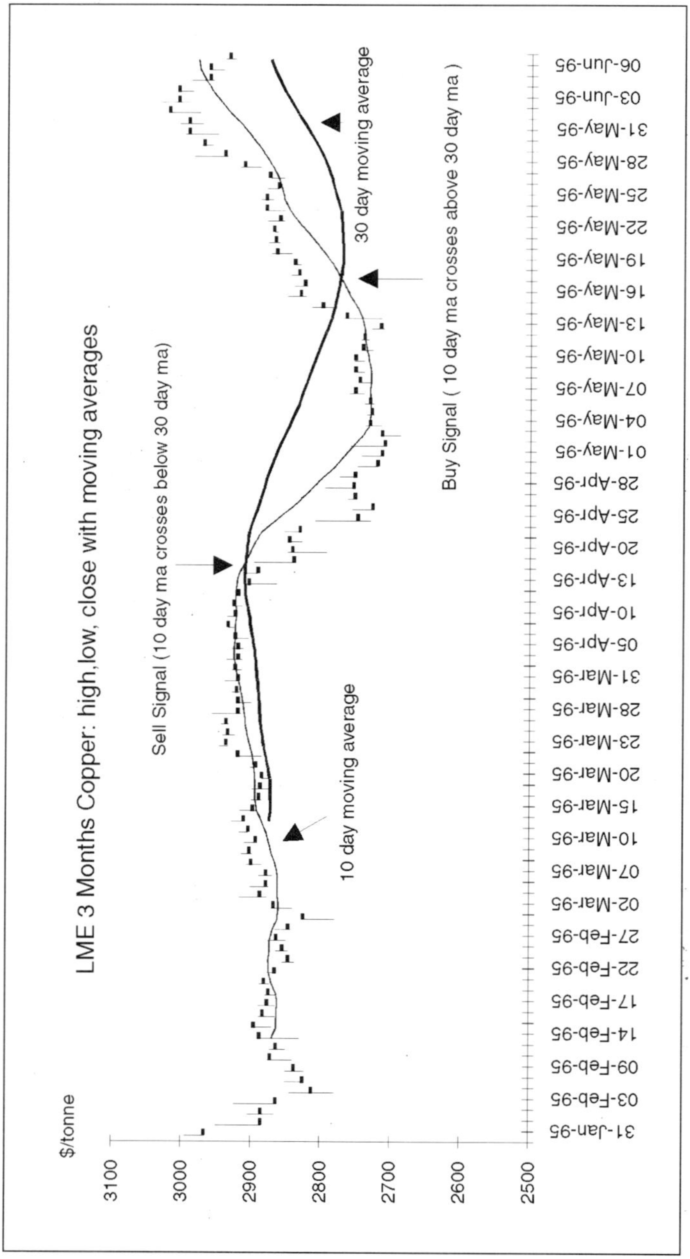
LME 3 Months Copper: high,low, close with moving averages
$/tonne
3100
3000
2900
2800
2700
2600
2500
Sell Signal (10 day ma crosses below 30 day ma)
Buy Signal (10 day ma crosses above 30 day ma)
30 day moving average
10 day moving average
31-Jan-95
03-Feb-95
09-Feb-95
14-Feb-95
17-Feb-95
22-Feb-95
27-Feb-95
02-Mar-95
07-Mar-95
10-Mar-95
15-Mar-95
20-Mar-95
23-Mar-95
28-Mar-95
31-Mar-95
05-Apr-95
10-Apr-95
13-Apr-95
20-Apr-95
25-Apr-95
28-Apr-95
01-May-95
04-May-95
07-May-95
10-May-95
13-May-95
16-May-95
19-May-95
22-May-95
25-May-95
28-May-95
31-May-95
03-Jun-95
06-Jun-95

Chart 3

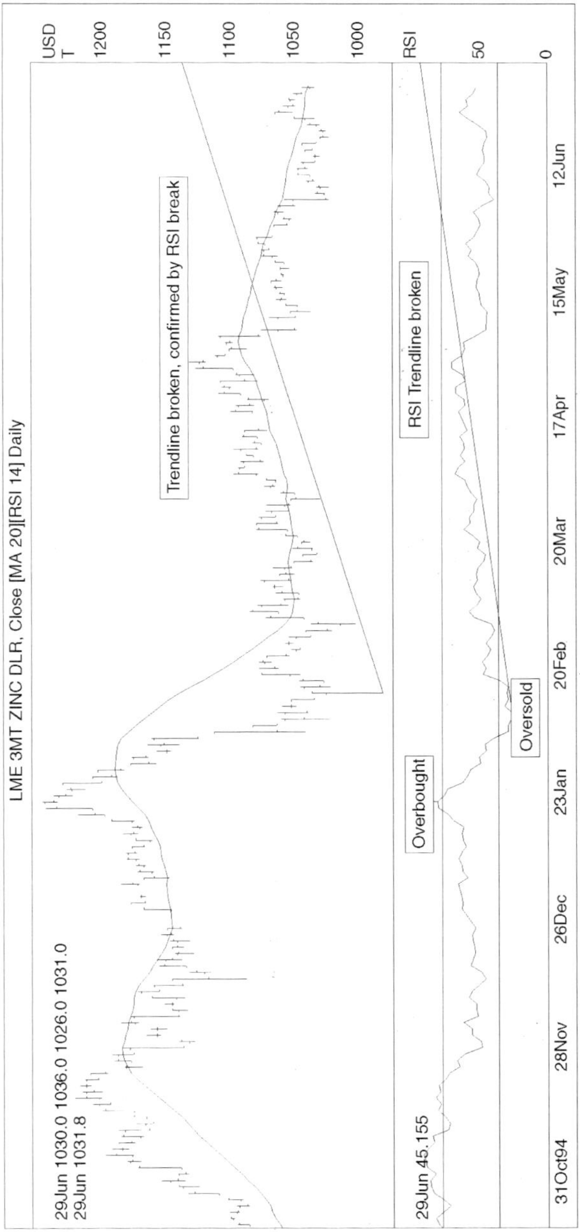
LME 3MT ZINC DLR, Close [MA 20][RSI 14] Daily
29Jun 1030.0 1036.0 1026.0 1031.0
29Jun 1031.8
USD
T
1200
1150
1100
1050
1000
Trendline broken, confirmed by RSI break
RSI
50
0
29Jun 45.155
RSI Trendline broken
Overbought
Oversold
31Oct94
28Nov
26Dec
23Jan
20Feb
20Mar
17Apr
15May
12Jun

Chart 4

CHAPTER 12

Aluminium

Aluminium is one of the most plentiful elements in the earth's crust, so it is not surprising that it has by far the largest tonnage production of any non-ferrous metal. In many of its markets it is in competition or in partnership with steel. Aluminium's key characteristics are lightness, corrosion resistance, electrical conductivity, ductility when pure and strength when alloyed. Its lightness explains its widespread use in all forms of transportation – aircraft, motor cars, ships and railways. Once considered a premium product, it is now as likely to be found in the body of a tipper truck as of a luxury car, or in the wheelhouse of a trawler as the cabin of an ocean liner.

Next in order of importance in its applications is construction, where it is used extensively as roofing, cladding and siding on factories, stadia, agricultural and residential buildings, often in prepainted form. As extrusions it is used in windows, doors and screens, often profiting from the metal's ability to take a durable, fade-resistant colour coating in an anodised finish. As tube, it is used in handrails, balustrades and scaffolding.

In the electrical field, aluminium is almost universally used in bare overhead conductor for the transmission of high tension current. Normally it is stranded on a core of galvanized steel or high-strength aluminium alloy to ensure cable can be strung in long lengths between towers. It is also used in insulated power cables, notably in industrial installations.

A rapidly expanding sector of use in recent years is packaging, where aluminium's traditional use as foil has been joined by aluminium cans, mostly to contain carbonated beverages. This application has benefited from environmental pressures. With their higher value and lower melting point than steel, aluminium cans have a strong case as an easily-recycled form of packaging. It is also the standard metal for the production of collapsible tubes.

It is important to understand metal flows in the aluminium industry to read supply-demand data correctly. For instance, a large part of the aluminium used in the automotive, white goods and engineering industries is as castings. But, as explained more fully in the chapter on aluminium alloy, castings are extensively made from secondary-origin alloy ingots and do not, therefore, directly impact the primary supply-demand balance.

PRODUCTION PICTURE

Aluminium production is a three-step process. It begins with mining bauxite, a reddish-brown aluminous earth in which the principal impurities are silica and iron oxide (hence the colour). Almost all extraction is by giant shovels operating in open pit mines, after removing overburden, the thickness of which varies from 2-3ft up to 40ft. Deposits are almost all found in tropical latitudes in the Caribbean, South America, Africa, South East Asia and Australia. The determinants of the competitiveness of a deposit are: grade of aluminium trioxide, content of impurities, water content (some deposits need drying before shipment), depth of overburden and accessibility to ocean shipment or adjacent

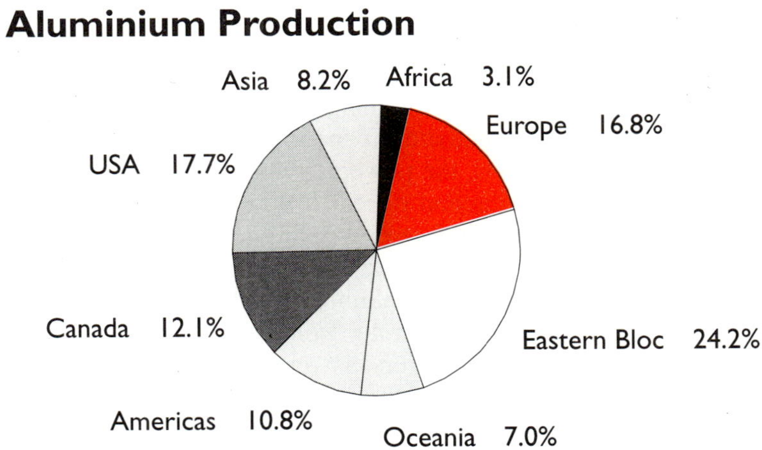

Source: Wolff Research

World production 18.6 million tonnes

Aluminium Consumption

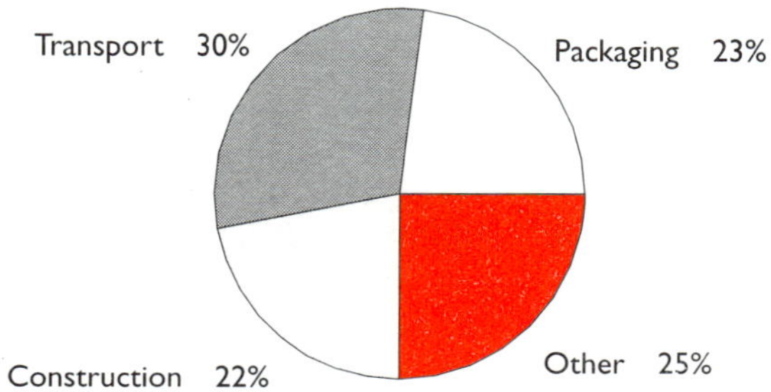

Source: Wolff Research

alumina refinery. With a fob value of only a few dollars per tonne, bauxite, and thus the mining phase of aluminium production, is not a big factor in market sensitivity. This is in contrast to the other LME-traded primary metals, which all have a mining-intensive psychology.

About two tonnes of bauxite are needed to produce one tonne of metallurgical grade alumina, the powdery, white intermediate product on which all aluminium production depends. When every other metal goes through the smelting phase before proceeding to refining, the aluminium industry has perversely borrowed the word refinery from the metallurgy of other metals to describe an alumina production plant, while the electrolytic reduction of aluminium from its oxide is described as smelting! Alumina is almost universally produced by the Bayer process; it was the development of this process along with the Hall-Heroult process for electrolytic reduction of the metal which formed the basis of the modern aluminium industry in the last years of the nineteenth century. It should not be forgotten that aluminium had a 60-year history in the nineteenth century as a chemically-produced metal with rising production and falling prices. However, it was the quantum leap in scale of production and reduction in costs associated with the Bayer/Hall-Heroult processes that marked the beginning of aluminium as the commodity we know today.

The Bayer process involves some use of energy to raise steam to heat the digesters in which the bauxite is dissolved in caustic soda. Alumina is then precipitated from the liquor by a seeding process.

Reduction of aluminium from its oxide is very power-intensive, which explains why significant parts of world production are clustered near hydro power resources in eastern and western North America, Latin America, Norway, Siberia, South East Asia and a few locations in Africa. The other economic sources of energy for aluminium production are low cost, often low-grade, coal and the off-gases from oil extraction. The latter is the sole basis of the fast-expanding Middle East industry. Oil itself rapidly disappeared from the picture following the "oil shock" of 1974. This also had an impact on ocean freight costs, and provoked a trend to move alumina plants closer to the bauxite mines to halve the volume of raw material shipped. At the same time the changed face of the geographical distribution of primary aluminium production sparked the development of a merchant market in alumina. Previously production of

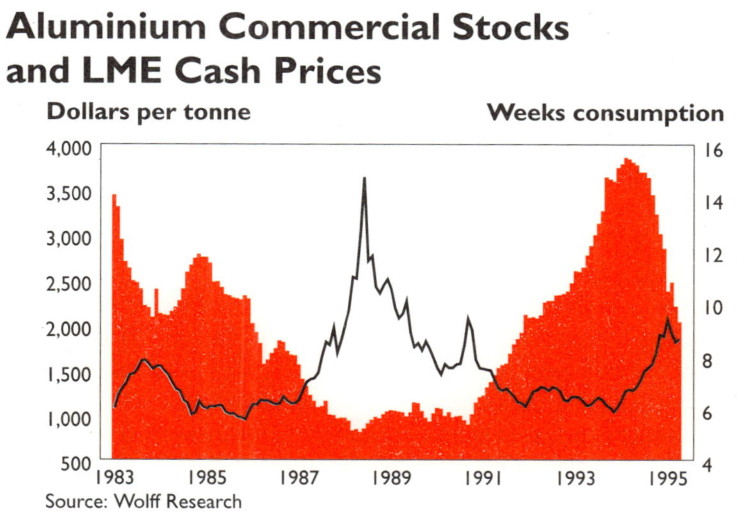

that material was confined to the major smelters' vertically-integrated operations. Even so, alumina production today is controlled by far fewer enterprises than smelt the metal.

Aluminium smelters are huge continuous electrolytic operations comprising cells, universally known as "pots" disposed in rows known as "potlines". A modern plant will have a capacity of not less than 100,000 tonnes a year, and perhaps as much as 500,000 tpy. A 100,000 tonnes plant will have perhaps 100 pots in just 4 potlines. The vast amounts of power arriving at the smelter are transformed and rectified to 1000v DC at 200-250,000 amps. Carbon blocks serve as anodes. They are prebaked from petroleum coke, though some plants still use the Soderberg process in which the carbon is fed continuously as paste into aluminium casings to form the anode. This is then baked in situ by the heat of the cell. The cathode is formed by the carbon lining of the cell. The heat generated by the passage of these large currents keeps the electrolyte molten. It is formed of a mixture of alumina and cryolite, a naturally-ocurring mineral now mostly substituted by synthetic cryolite made from fluorspar. As electrolysis proceeds, molten aluminium is formed at the bottom of the cell. Fresh alumina is continuously fed into the cell, while once every 24 or 48 hours the accumulated aluminium is syphoned off and taken to the cast house. It is already at commercial purity of 99.7%.

Modern environmental awareness has influenced changes in cell design to assure the collection and treatment of the fume from the electrolysis process, notably its noxious fluorine content. Cells are mostly hooded and the collected gases are scrubbed. Often, as a first stage, this is done by passing the gas through the incoming alumina, thereby recovering fluorine values.

Anodes are mounted on racks to allow their depth of penetration into the molten bath to be continuously monitored and adjusted, as this has an important bearing on the efficient use of the electric power. When anodes are spent, the complete rack can be removed and a new one substituted with the minimum interruption to production. Cell linings should last many months. When they need renewal, the complete cell is taken out of service, the busbars collecting the current are cut and the whole cell either lifted out by one or two cranes, or, in the most modern plants, lowered into the cellars, where it can be replaced by a freshly-lined cell which is then hoisted into position. A cell complete with its lining is very heavy, and in some older plants the spent lining has first to be laboriously chipped out with jack hammers to bring the bare cell weight within the capacity of the cranes.

All this explains why the problem most feared by smelter managers is a loss of power or a sudden strike of workers. Once the routine of production is interrupted, cells are liable to freeze – ie solidify – and rectifying that situation takes many months of heavy labour. Because the process is electrolytic, simply switching the power on again will not re-melt the electrolyte – let alone the aluminium.

SEMI-FABRICATION

Aluminium is fabricated by all the conventional means – rolling, extrusion, drawing and forging – used in other non-ferrous metals. It also has a few extra tricks up its sleeve more normally associated with steel – for example, the production of welded tube from strip. As rolled product, it is sold in many thicknesses. These start with a 6″ thick slab of high duty alloy which, after ultrasonic testing, goes to be formed and skin milled into part of an aircraft fuselage. At the other extreme, it is rolled down to foil whose thickness is measured in microns, and which may be used in cigarette packets or, after suitable etching, in condensers. Aluminium is the most easily extruded structural metal, and the applications of extrusions are legion. Glazing bars, as used in roofs of factories, conservatories and public buildings are an excellent example of the metal's ability to produce fiendishly complex cross sections with ease. Simple forms like round rod for machine shops may be produced a dozen at a time. Travellers by air have good reason to be thankful for the strength of aluminium forgings, as this is how the main structure of aircraft landing gear is made. There are many other applications in high performance and defence markets.

Aluminium wire is almost all produced by continuous casting and rolling prior to final drawing and by far the bulk of production is in EC (electrical conductor) grade for bare overhead transmission lines. Some wire is produced in harder alloys for mechanical applications, and small-diameter pure wire has a use, often when anodised, for decorative purposes.

PRICE FORMATION

The "big six" aluminium smelting companies (who have some similarities with the "seven sisters" of the oil industry) have tended over the last fifty years to blow hot and cold on the virtues of vertical integration. From a background in the thirties and forties of extensive vertical integration, both upstream to alumina and bauxite production and downstream to semi-fabrication and even finished products, they have progressed unevenly to their present position, which emphasises primary metal production. All the same, an important part of the world's sheet rolling capacity is still controlled by the majors. At the other extreme, aluminium extrusion and wire and cable production are almost entirely in the hands of independents. This evolving business philosophy is interwoven with the history of moves towards bringing aluminium on to the LME and with the development of that contract since it was first traded nearly two decades ago.

In the fifties, sixties and seventies, the aluminium industry worldwide believed passionately in the merits of producer-determined prices as the correct mechanism for commerce in its metal. When more than half the sales by a smelter to a semifabricator were in effect an internal transfer within the same ownership, the origins of this approach are understandable, if not exactly popular with taxation authorities or classic economists. During this period, several external forces combined to transfer fabricating capacity from the big six smelters to independents, quite apart from the big smelters' own periodic doubts about the wisdom of being so heavily in downstream activities. Extrusions and cable have already been mentioned, while the development of continuous strip casting somewhat moved the goalposts on strip and sheet rolling.

HOW THE CONTRACT HAS EVOLVED

Nevertheless, when aluminium trading on the LME was first seriously mooted in the mid-seventies, even the independent fabricators joined with the smelters in asking "who needs it?" It was only the introduction of trading as something of an act of faith in 1978 that began to produce answers to that question. Independent ingot traders soon realised the value of being able to hedge their metal positions, while independent smelters, notably those in the then-USSR, appreciated that putting metal into LME warehouse not only solved their problems of marketing the material, but produced prompt cash receipts in hard currency. However, before this happened, the Exchange had to learn some facts of life about the differences between aluminium and the older metals it was accustomed to trading.

On the model of the other contracts, aluminium was introduced on the basis of metal being deliverable in 12-26 kilo ingots, but it was not until a few years had passed that it was realised that this form was unduly restrictive on the deliverability of metal to a trade-oriented market, and

T-bars were added. These are a large, approximately 600kg shape, the fat T-shaped cross-section of which facilitates their handling by fork lift truck and stacking. It also says something about the climate of industry hostility to the contract at its inception that deliverable brands were identified by listing by the exchange, rather than by their producers registering them. Subsequently sows, an approximately 3/4-tonne lump of metal, were added. However, it was the change in the deliverable grade from 99.5% to 99.7%, and contingent changes in the Fe and Si limits that decisively opened the market up to deliveries of USSR material. It also looks rather quaint, with today's hindsight, that the contract was introduced in £ sterling rather than US dollars. It is indicative of the need for evolution as well as revolution in establishing a futures contract that attempts by the Comex in recent years to introduce an aluminium contract have been decidedly unsuccessful.

Even today, there are still those in the industry who believe that aluminium prices are set by the producers, and that the LME prices are some kind of optional extra commentary on the producer prices. In fact, this situation began to emerge as much as forty years ago, when Metal Bulletin introduced a "free market" quotation alongside the producer price. The reality for the last ten years at least is that the LME aluminium contract is the leading, and for all practical purposes sole mechanism for discovering the world price of the commodity. This is reflected in the contract consistently having the largest volume of turnover of any base metal futures contract, as well as in the huge volumes of stocks – peaking at 2.5 million tonnes in 1994 – that the Exchange's warehouses were able to accommodate.

This dominant position of LME prices as the industry's reference price applies not only downstream to semis prices, but also upstream where contracts for alumina supply are often expressed as a percentage of the prevailing aluminium price, thus affording independent alumina producers and traders, too, scope for achieving some hedge protection.

Today, as with older contracts like copper, a virtuous circle is in place whereby the LME, through its prices, signals implacable, if sometimes unpopular, facts of life to the physical industry, which in turn adapts to those facts more promptly than it might otherwise have done. Perhaps the best evidence of this is the "Memorandum of Understanding" formed between all the major world producers in 1993 that production cuts were necessary to reduce the massive overhang of unsold stocks, some 2.5m tonnes of which were highly visible in LME warehouses. Having taken the precaution of involving their governments in these talks, the producers were able to implement the smelter closures they agreed were necessary without immediately being arraigned for operating a cartel. The impact of the closures on both stocks and prices was prompt and predictable and is continuing as this book goes to press. However, the US authorities, despite their involvement in the original talks that led to the Memorandum, have since mumbled that an investigation into possible monopolistic practices should be made. The reaction of market sentiment to the doubtless on-going and changing news on this whole front can be expected to be a market feature for a number of months yet.

CHAPTER 13

Copper

The dog is often described as man's best friend, but there are good reasons for suggesting that copper has more claim to the title. Copper was the first mineral that man extracted from the earth and converted to a specification material. With a little help from tin, it was the reason for the Bronze Age; and it could be argued that without the Bronze Age, we would still be in the Stone Age!

Ironically, Bronze Age man could hardly have foreseen that copper's critical value to industrialised society today owes almost nothing to bronze and almost everything to electricity. Beginning with the dry cell-powered railroad telegraphs of the mid-nineteenth century, the symbiotic relationship between electricity and copper has never looked back. Most of today's use of electricity, from the humblest Walkman to the mightiest electric arc furnace depends at some point on the delivery of that energy through flexible, insulated cables. And thanks to its electrical conductivity, only a short step behind that of the expensive No 1, silver, copper is the standard material for this application.

Additionally, copper's thermal conductivity is vital to the performance of many heating and cooling devices, from a skyscraper's air conditioning system to the radiator of the family runabout. Its corrosion resistance is, again, among the best of the base metals, as is attested by Bronze Age coin finds or cannon recovered from men o' war that sank over 300 years ago. This property is routinely relied on in a myriad of modern applications, which in turn rely on copper's huge range of alloys, developed over the millenia. These can be tuned for ductility (pure copper), strength (various bronzes such as gunmetal, manganese bronze, silicon bronze), lubricity (bearing materials like gunmetal, phosphor bronze and aluminium bronze), machinability (leaded brass), formability (60/40 brass for hot die stampings) and appearance (from gilding metal for lipstick cases to a patinated copper roof). Additionally, copper is one of the most easily-joined industrial metals by soldering or silver soldering.

All these advantages for modern service conditions explain why man's oldest metallic friend is still alive and kicking. Indeed, it enjoys a secular growth rate in consumption, at about 2.75% a year, which is the envy of several other base metals and steel. However, change is a constant process, and copper is also under attack in a number of its applications. In telecommunications, the battle with fibre optic cable has effectively been lost, as far as trunk lines are concerned, though the explosive growth worldwide in the installation of telephone instruments is itself a consumer of copper, apart from requiring internal distribution wiring in copper in the home, factory or office. The automobile radiator, at one time a major consumer of copper, brass and even nickel silver, is increasingly giving way to the lighter aluminium unit, now that the jointing problems of aluminium have been overcome.

In the mid-range, insulated power cable market, the battle between copper and aluminium is still in full course, influenced by non-technological factors like the relative medium-term price levels of the two metals, the flexibility of building codes and even the prejudices of

Copper Production

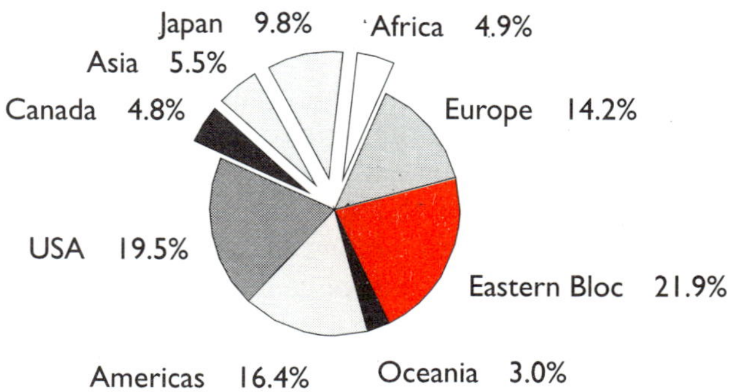

Source: Wolff Research

World production 11.4 million tonnes

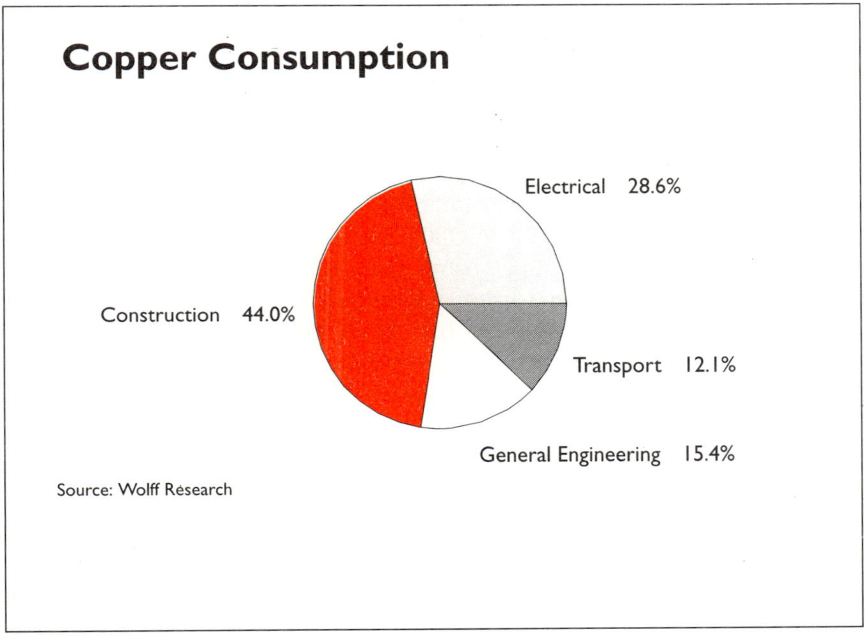

designers and installers. Only in the field of water tubing for heating and air conditioning does copper seem reasonably secure in its present applications and at its present relative price level. It was not always so. Back in the sixties, when copper suffered perennial shortages, much effort was expended on its substitution by stainless or coated steel.

RESOURCES

Copper is not a particularly rare metal in the earth's crust, but the reserve of known, economic deposits is not so generous that it could compare with, for instance, aluminium or steel in this regard. As a result, it is a market which is considerably influenced by mining considerations. A situation such as is found in the USA, for example, where not only mines but also smelters widely employ workers from the same union, makes the market very sensitive to impending labour negotiations, or the actual outbreak of strike action. Similarly, the location of many producing operations in areas of actual or potential political instability is another fruitful source of market anxiety. Over time, developments such as the civil war in Zaire, local insurgency in Papua New Guinea, political upheaval in Chile or terrorism in Peru have demonstrated that this is a real factor in the copper supply-demand picture.

At the same time, it is a little strange that copper is so obsessed with these factors related to primary production. Apart from lead, copper has the highest percentage of supply originating from secondary sources of all the base metals – around 25% worldwide, a good deal higher in developed markets. Moreover, this supply stream is more fully integrated into the total supply picture than in any metal other than lead. This is because an important proportion of copper scrap goes for refining and returns, as cathodes or ingot bars, in forms indistinguishable from their primary counterparts. Additionally, alloys like brass have a large recirculating scrap component into which primary metal hardly penetrates.

By the same token, the market is rarely moved by developments that actually or potentially affect first-line consumption, which is influenced by both consumer and capital spending all around the world. Because of its longevity as an Exchange-traded metal (copper was the first metal traded when the Exchange opened) and supported by a copper futures market in the USA since the thirties, the market is totally focussed on the LME price as the arbiter of where the market is standing. Thus everybody,

Copper Commercial Stocks and LME Cash Prices

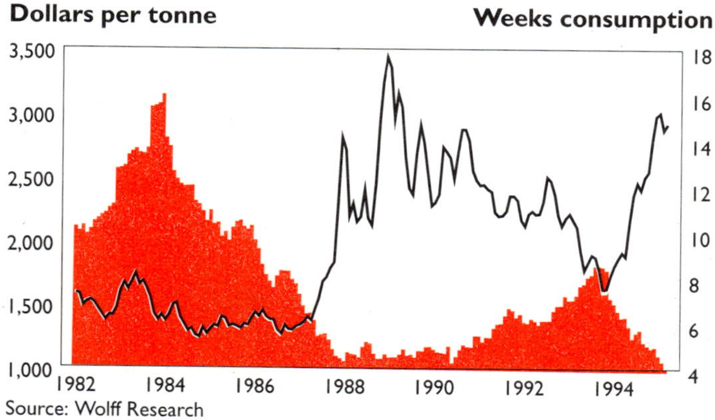

Source: Wolff Research

from the geologist exploring for copper in the field down to the street corner scrap dealer, focusses on the LME price and is considerably influenced by it. It is for all these historical and, as it were, cultural reasons that copper is referred to as the LME's "flagship" contract. This in turn explains why events in copper will influence sentiment in the other LME metals, even though the events themselves may have no relevance to the other metals. The compliment is sometimes returned and other metals influence sentiment in copper, but this is not so common.

PRODUCTION

Primary copper production, as indicated, begins with mining, but then, mainly under the influence of the type of ore extracted, splits into two streams, the technology of which should be understood, at least in outline, by followers of the market. On the one hand, ore as-mined is crushed, concentrated by flotation, pyrometallurgically smelted to blister in a reverberatory or flash furnace, fire-refined to anode in a convertor and finally electrolytically refined to cathode. This is the classic route, and usually applies to sulphide deposits.

Alternatively, the ore as-mined is heaped up on a pad with an inclined impermeable base and sprinkled with acid. This leaches out the copper content (in which process assistance may be given by bacterial action), the leach liquor is collected and becomes the electrolyte in an electowinning operation. The process is known as Solvent Extraction - ElectroWinning or SX-EW for short. The growth in importance of this process over the past two decades has been little short of phenomenal. It is mainly applied to oxidised ore. As it is not uncommon to find a sulphide deposit whose upper layers have been oxidised, especially in the large, disseminated low-grade deposits that are now being considered economic, it is also quite common to find both routes being used at the one deposit.

An elegant feature is that it is the oxidised cap of the deposit which is extracted first, and it is this that requires the lower capital cost SX-EW installation. This can then begin to produce a revenue stream, where previously the oxidised ore had to be considered merely as part of the overburden. Another advantage of the SX-EW process for deposits in remote locations is that it is something of a recirculating method, in that the spent electrolyte is a major input to the leach acid make-up. In the early days of SX-EW, the cathodes produced suffered from a conventional view that electrowon metal cannot be as pure as that which is produced electrolytically. (The distinction between electrolysis and electrowinning is that the former starts with copper metal at the anode, and this is progressively consumed as the pure copper is plated out at the cathode; in electrowinning, the anode is inert and the copper is recovered from the electrolyte). This view was supported by the experience that the only electrowon copper known to the industry in earlier times came from Zaire, and the state of technology at the time those operations were set up in the twenties required the electrowon cathodes to be finally fire-refined in the process of re-casting into wirebar form. Today, it is accepted that, barring geological quirks affecting control of the leach process, electrowon cathodes are of as good quality as those produced by electrolysis.

FABRICATION

The reference above to wirebars recalls a troubled period in the Exchange's relationship with the physical industry which is now virtually at an end. Historically, the classic method of converting copper to its most-used form as electric wire involved casting wirebars (a shape of approximately 6″ square cross-section weighing about 120 kilos), which were then hot-rolled in a multi-stand mill to copper wire rod of, normally, 8mm diameter. Over about the last 30 years, various proprietary processes of Continuous Casting and Rolling have been developed which have now obsoleted wirebar rolling. In the CCR process cathodes are melted in inert atmosphere in a shaft furnace and the molten metal fed continuously to a continuous casting installation which produces a square or irregularly shaped rod which is immediately rolled in a multi-strand mill, synchronised with the output of the caster, and producing wire rod. The advantage of the CCR process is better control of the metal quality during melting, somewhat better metallurgical quality of the rolled rod and the ability to produce coils of rod of considerable length and weight. For the wire drawer, this avoids the need to keep joining up the shorter coils to which the wirebar is limited, with the attendant risks of failure at this point in subsequent drawing. At the time of intense rivalry between the two production processes, champions of each fought for equal, or if possible better, terms of deliverability on to the Exchange of their wirebars and cathodes respectively.

Other types of copper fabrication are more consistent and conventional. Tube is produced by casting billets, converting these to tube shells by hot piercing or back extrusion and rough drawing. Thereafter, tube can be drawn cold, with intermediate anneals, in considerable lengths thanks to the use of a floating mandrel. Water tube is then cut to standard lengths, internally cleaned and tested for pinhole defects hydrostatically or with eddy current detectors. Sheet is produced from rolling slabs (which can also be continuous-cast) by hot and cold rolling. Top quality rolling, especially of brass, sees the hot rolled plate skin-milled both sides to remove the oxidised surface. Thereafter cold rolling and annealing proceeds in the usual way. Brass was, of course, the material with which the extrusion process was first applied to the production of finished shapes such as rod, channel, angle etc. (Today aluminium has far overtaken copper alloys in volume of production). Standard brass extrusions are round, square and hexagon rod, both solid and hollow, for use in machine shops. A small part of production goes to hot stampers making such parts as valves and faucets, though there is now competition from certain resin mouldings in this area.

INDUSTRY STRUCTURE

Historically, copper has never had the degree of vertical integration of, for example, the aluminium industry, even in the USA, where vertical integration was certainly much more in evidence than in Europe in the fifties and sixties. Today, there is a reasonable amount of vertical integration between mining, smelting and refining, but relatively little forward from there to fabrication, and very little from semi-fabrication to end-product manufacture, except for some instances of integration of cable manufacturers with producers of industrial transformers, electric motors and generators. Even at the mining-smelting end of the chain, there is a good deal of arm's length trade in concentrates, blister and cathode. A major factor in this structural feature is the custom smelters who buy both blister and secondary refining materials and sell their cathode or fire-refined output to independent fabricators. Despite their apparently more precarious position than the smelters vertically-integrated with their mine supplier, they remain a significant force in the market, especially in Europe and the Far East; there is no doubt that their ability to shop around flexibly between primary and secondary sources of raw material is part of their secret.

An aspect of copper production that should never be forgotten is that it is a rich source of by-product revenues from materials such as gold, silver, germanium, selenium, and tellurium. These materials are usually recovered at the electro refining stage, but of course have to be paid for where assay indicates their presence in the blister or scrap to a meaningful extent. With secondary raw materials, there are also often tin and nickel values to be recovered. Together with the considerable volume of copper alloy scrap that circulates without ever reaching the refiners, it is easy to understand why statistical models of copper supply and demand are more complicated than those for other metals.

PRICE FORMATION

In view of the LME's dominance of the world copper pricing scene, and of the additional presence, especially in North America, of Comex prices, it is not surprising that the copper reference price situation is quite simple. If you want to know what the price of copper is, you start with the latest LME price and adjust for the particular features of the parcel you are considering. These adjustments may be for grade, shape, brand, distance from LME warehouse, and date of availability, to name only the most obvious. One feature that enlivens LME Dinner week each October is the ceremonial announcement of the Zambian and Chilean producers' terms for their sales to contract customers during the following year. Expressed as a premium over their producer price (which is simply the LME settlement price on the date of transaction), these terms are around $20 per tonne, less for pricing on the average and more for pricing on the unknown. After many years of tussling between these producers and independent fabricators, the longstanding facility of "backpricing" has finally disappeared, thus shortening this edition of this book by several hundred words compared with its predecessors!

Recently, much has been made of supposed conflict between the LME and Comex for domination of physical copper pricing in the US market. On its side, the LME has finally extended its worldwide network of copper-carrying warehouses to the USA. Previously its US warehouses had been limited to its other five traded primary metals, as there is no meaningful US futures market for those metals. At the same time the Commodity Exchange Inc in New York (Comex), which had become rather small by modern futures market standards, finally merged into the more powerful New York Mercantile Exchange (Nymex). This has led to several moves to reassert the status and reinvigorate the activity of Comex, notably in its second most important contract, copper.

Inevitably this has made the LME warehouse move appear hostile; conversely, so does Comex's declared intention to introduce an effective contract in aluminium, a metal it had previously tried without success to add to its range. In principle, this competition is good for exchange users and even price-takers. But to take most advantage of it does require them to understand the different operating systems of the respective exchanges, and the differences in price-taking practices between the market in the USA and that outside the USA.

CHAPTER 14

Lead

For some years past it has been fashionable to decry lead as an old-fashioned metal with one application – motor vehicle starter batteries – and a poor environmental status. So it may come as a surprise to find that lead's secular consumption growth rate, though modest in comparison with, for example, aluminium, is steady and positive. In parallel with the dominant place in lead end-uses of the motor car battery, the love affair of the human race with motor cars is largely responsible for this performance. For it must be remembered that it is not the rate of production of new cars which principally determines the level of demand for batteries, but the number of cars in operation on the road – the parc as it is called. And this keeps rising worldwide.

This dominant end-use also goes much of the way to explaining the high proportion of total lead consumption coming from secondary sources – around 60% in developed economies. This is despite the actually and potentially negative impact on the well-established lead recycling business from well-meaning environmental legislation and regulations coming from national and supranational bodies. A scrap automobile battery is, if handled correctly, a neat non-polluting package in which its lead values, and sometimes some of its plastic values, can be transported to an environmentally-sound operation for processing. The recycled lead is, of course, already formulated for further use in batteries.

ENVIRONMENTAL PROS AND CONS

Even more ironically, bearing in mind that environmentalists are agreed in seeing lead as a dirty word, the lead-acid battery is accepted as the most likely stepping stone to the new generation of so-called zero-emission vehicles. The electric town cars that are envisaged by this programme may, one day, be powered by a battery of another system than lead-acid. But any legislation designed to bring these vehicles on to the streets within the next five years will almost certainly force the first generation to be powered by the proven lead-acid battery. This would, of course, initially be incremental demand on top of existing demand for starter batteries, and would therefore have to be mainly supplied from mine-produced metal – and in a ratio of much more lead per vehicle than at present.

As dominant in lead consumption as the lead-acid battery is, it is nevertheless complemented by a variety of other uses that draw on lead's special combination of corrosion-resistance, softness, low melting point and high atomic weight. Most traditional of these applications is lead's use in solders and low melting point alloys, while the architectural use of lead sheet has enjoyed a small revival recently. Much more modern are its use in pvc stabilisers and radioactivity shielding; this latter may range from hundreds of tonnes in a lead brick wall in a nuclear power station to a few grammes of lead foil backing a dental X-ray film. Lead's softness gives it value as a compact and efficient sound barrier (environmentally satisfactory for those benefiting from the peace!). Before writing lead off as a metal with only a limited future, it is as well to recall the uses which it has lost in the past, but still survived. These include lead oxide in paints

Lead Production

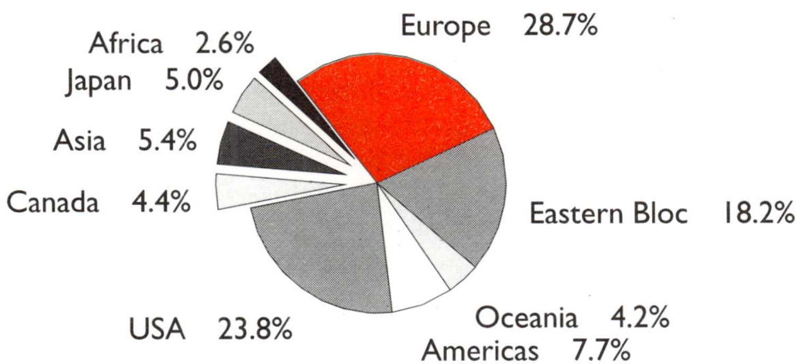

Source: Wolff Research

World production 5.4 million tonnes

Lead Consumption

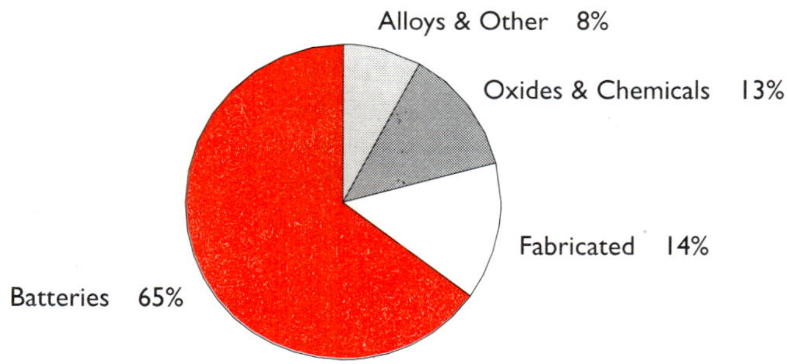

Source: Wolff Research

(the biggest prewar use), lead pipe in plumbing, lead sheathing for cables, lead typemetal and tetraethyl lead in petrol; this last has now been legislated off the market in certain countries, and can be expected to be banned from more.

In the same way, the survival of the lead-acid battery as the universal means of starting internal combustion engines is not solely due to the unfulfilled promise of competing battery technology. It is also due to the lead-acid battery's own evolution in terms of power-weight ratios and effective life. Additionally, revision of the alloying elements used to make battery lead has given us the so-called "maintenance-free" battery, in which the traditional chore of topping up the electrolyte with distilled water has been eliminated. Briefly, this has involved the reduction or elimination of the antimony content of the alloy in favour of a certain amount of calcium, tin and/or arsenic. In assessing the importance of the battery to the total lead economy, it should be remembered that not only are the plates and grids made of lead, but so, too, is the lead oxide paste which is the electrically active element in the positive plates. When recycled, the lead content of this paste is reduced from the oxide, as well as the melting of the plates. Looking to the future, it would be foolish to predict that the lead-acid battery has exhausted all its possibilities for further development.

PRODUCTION

The mined lead that makes up about half total supply comes from a wide geographical spread of countries, notably the USA, Russia, Australia, Canada, Mexico and Peru. With its relatively simple, low-temperature metallurgy, smelting and refining operations are equally widely spread. There is, however, an active international trade in concentrates, though not as influential as is this element in zinc. Lead, zinc and silver are commonly found together in nature, and any one or two may be produced as by-products or co-products of the third. This tends to make producers' response to price movements less precise than Adam Smith would have expected when he wrote the law of supply and demand. Mining is mostly underground classic vein mining of galena followed by differential flotation to separate the lead from the zinc. Concentrates are roasted, then smelted in a blast furnace to produce bullion of 98-99% purity. Much work is going on to introduce turbulent layer roasting to lead with the

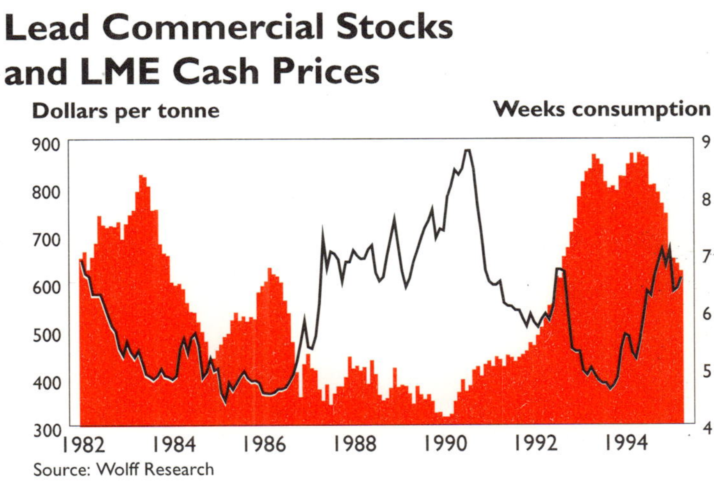

same success that has applied to copper. However, this has suffered some well-publicised setbacks in recent years.

Lead is finally fire-refined to commercial purity of 99.97% or 99.99%. The latter quality usually commands a premium for physical delivery, and as a result, brands of this grade may be hard to find in LME stocks, even though they are deliverable. Lead refiners can benefit from a range of by-product credits. Apart from the silver, which stays with the lead up to the bullion stage, these may include antimony, bismuth, indium, arsenic, gold, copper and tin. The values of these by-products can be quite appreciable, especially if they are refined up to five nines quality. Smelters and refiners located in developed economies are usually able to handle both primary and secondary raw materials.

PRICING PRACTICE

Outside the USA, the use of the LME price as the basic reference is almost universal. Somewhat more susceptible to import duty considerations than, say, copper, lead has many local market values derived from the core LME in-warehouse price. Additionally, because of its status as the lowest-value LME metal, the influence of elements such as duty, freight from LME warehouse base point and currency conversion is proportionately greater than for other metals. In the USA, a posted producer price system is dominant, but it is still ultimately linked to the LME through physical arbitrage and the deliverability on the LME of all the significant US (and Canadian) brands. Because of the ease with which secondary lead materials can be refined to a grade indistinguishable from primary, lead suffers less than the other base metals from the traditional problems of hedging scrap and residues on the futures market. Movement of scrap prices proportionately in step with the LME is more likely, but cannot, of course, be guaranteed.

CHAPTER 15

Nickel

Nickel is a metal of several special characteristics of considerable value in a high-tech world. For the most part, they are applied by alloying nickel with other metals, notably ferrous metals. This dependence on a branch of the iron and steel industry for an important part of its demand picture is not peculiar to nickel; zinc and tin are both in somewhat the same position. But whereas galvanized steel and tinplate are single commodities that dominate zinc and tin demand respectively, nickel is much more versatile.

It is true that stainless steel is the largest single outlet for nickel, taking just over half total consumption. But the impact of this fact on the LME nickel contract is not straightforward. This is because nickel production is classified in two distinct groups, usually known as Class I and Class II materials. The nickel forms deliverable against the LME contract are all from Class I, which means Class II materials are relatively much less directly influential on, or affected by, LME price movements. Obviously, at the end of the day the industry adapts its nickel consumption patterns to use the most cost-efficient form, regardless whether it is Class I or Class II. But this is a process much constricted by technology and conservative trade practice, so there are often times when quite a wide "arbitrage" opens up between the value of nickel units in Class I and Class II forms. But as there is no futures market in Class II, it is not possible to profit from this "arbitrage" except by complex and risky swaps in the physical market.

Indeed, nickel is generally one of the clearest examples among LME metals of how a little knowledge can be a dangerous thing. To a steelmaker in particular, the Class I forms deliverable under the LME contract – cut cathode, pellet and briquettes – may well be his raw materials of second choice. There are two, or even three, reasons for this. Firstly, Class I products are, with very minor exceptions, the only forms that can be used in non-ferrous and nickel-based metallurgy, but can also be used in ferrous metallurgy. So they are more valuable per unit of nickel than Class II products, which are restricted to ferrous metallurgy. Secondly, Class II products can bring additional benefits to the steelmaker by the nature of their composition. Thus ferro-nickel, which contains 25-30% Ni, balance iron, brings the steelmaker 2-3 units of melting iron for every unit of nickel supplied. Ferro makers describe this as "free" iron in their sales patter, but it is, of course, no more free than a free lunch. Nickel oxide has a valuable cooling effect when it is added to molten steel. Thirdly, at least in developed markets like North America, Europe and Japan, steelmakers will certainly be on the lookout for various kinds of nickel-containing scrap and especially, if they are making stainless steel, for stainless steel scrap.

So it can be seen that caution needs to be applied to simplistic analyses like "Stainless steel production is up 10% this year, therefore LME nickel prices will rise 10%". Just to underline this point, whilst all stainless steel contains chromium, not all stainless steel contains nickel. Other applications of nickel in ferrous metallurgy also exploit nickel's corrosion resistance. Its ability to confer this on that most corrodable of metals, iron,

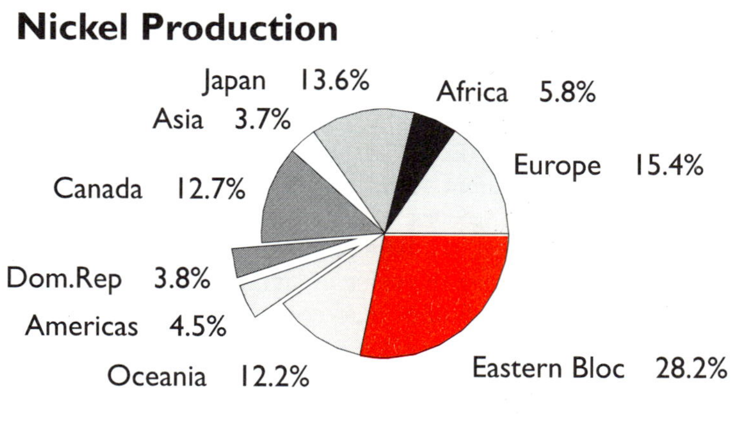

Source: Wolff Research
World production 825,000 tonnes

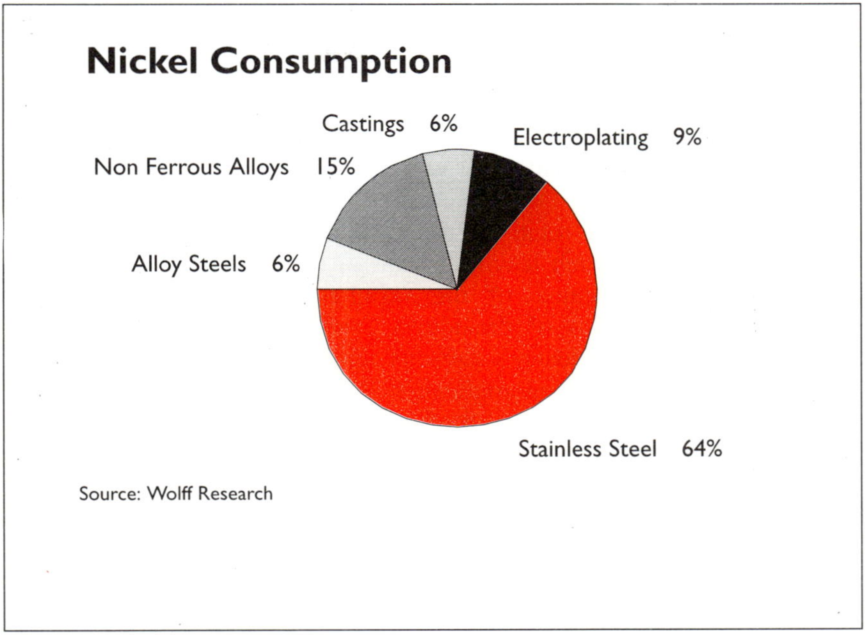

is found in complex engineering alloys of steel or iron. In the latter case, nickel may well also be conferring on the alloy some of its own high heat resistance. Indeed, without going to noble metals like platinum or exotics like tantalum, nickel is the pre-eminent base metal for corrosion and heat resistance in all kinds of aggressive environments. Last but not least of the ferrous materials are the heat treatable engineering steels, without which all machinery would still have moving parts the size of a Victorian steam engine. Nickel is particularly valuable in conferring strength and toughness (resistance to brittle failure) on heat treated steels.

On the non-ferrous side, nickel is found in conjunction with copper (nickel silver, cupro-nickel and Monel), aluminium (in ships' propellers), tin (as an electroplating alloy) and as the base metal in a vast range of high performance alloys. Among high profile applications of such alloys are aero engine parts, often made in the Nimonic series of alloys. Less glamorous, but just as important for tonnage offtake, are applications in chemical and nuclear plants, glass-making and ceramics. As befits a metal closely associated with technological change, nickel's patterns of use are themselves consantly changing. What with changes in fashion and increasingly stringent functional demands on components, a large part of the market for nickel as a substrate to chromium in plating has disappeared. Changes in telephone equipment design have wiped out the demand for nickel silver contacts. The one-time standby of the sound recording industry, the vinyl record has given way to tape and the CD; with it has gone the application of nickel in electro-deposited masters of the records.

Nickel semis are a relatively obscure part of the industry, partly because they are so much less important in tonnage terms than ferrous alloys and partly because of the sophistication of the range and performance capabilities of the alloys used. This has resulted in an important part of the industry's capacity remaining vertically-integrated. Nickel also has a flourishing application in chemicals – known as nickel salts – and as a catalyst. One of the best-known such applications of the latter is as Rainey nickel, used in the manufacture of edible oils.

PRODUCTION

If the demand picture for nickel is complicated, the way this is interwoven with production and the standard forms in which nickel is sold only

Nickel Commercial Stocks and LME Cash Prices

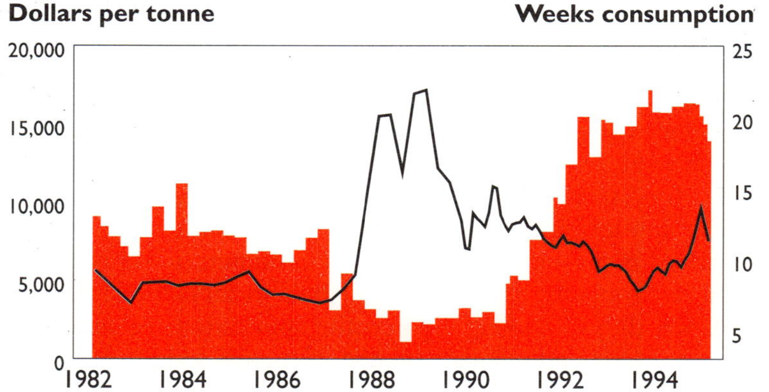

Source: Wolff Research

complicates matters further. We have seen how the lower-value Class II products are suitable only for alloying with iron and steel. This is because they come from laterite oxide deposits or, uniquely in New Caledonia, from garnierite, a nickel silicate. As open pit operations, oxide deposits are usually cheaper to mine, but processing to oxide or ferro-nickel is energy-intensive. New Caledonian nickel is the oldest source in the industry, but oxide deposits such as are being exploited in Cuba, the most senior oxide producer, Indonesia, the Philippines and Brazil have experienced varying fortunes in the past forty years. To rub salt in the wound, not only do sulphide deposits yield Class I products, they also often enjoy considerable by-product credits from copper, platinum group metals, cobalt and other minor metals. This is particularly true of the two major Northern hemisphere sources, the Sudbury basin in Canada and Norilsk in Russia. Again, whilst oxide ore naturally yields two forms of nickel which are often mutually competitive, sulphide ore uses three processes to produce nickel. Each of the forms thus produced has certain properties that distinguish it from all the others.

Pellet is produced by a version of the original Mond refining process, which goes via the highly poisonous gas nickel carbonyl. The product is among the purest, in a form resembling ball bearings; it is not, therefore, a material to be spilled on the foundry floor. On the other hand, it lends itself to automated feeding processes, and unlike cathode, does not have to be further processed to be easily handled. Cathode is the most universally traded form of raw nickel as it dates back to the days when pellet was an Inco monopoly reserved for direct consumer sales and briquettes had not been invented. It is also the form in which Russian nickel is sold. Full plate cathodes, commonly a metre square, are normally sold cut to 10″ x 10″ down to 1″ x 1″ squares, when they are suitable for a wider range of consuming processes than any other form. Purity is normally up with that of pellet. Briquettes are compressed from powder produced hydrometallurgically by the Sherritt process. They are pillow-shaped, usually 1.5-2″ long. They are not suitable for plating or any application where there is a risk they will crumble during processing. The whole point of making briquettes is to avoid the losses during melting associated with powder. However, some powder is sold prior to briquetting, notably for powder metallurgy.

Excluding Class II products from what is deliverable on the LME makes for a less geographically balanced nickel supply picture than obtains when taking Classes I and II together. On the other hand, it was at first considered that, even after limiting the LME contract to Class I forms, nickel was impossible to "commoditise" because of the variety of forms in commercial use. The successful selection of contract terms, not only for grade but also for other aspects like warrant size, warehouse control of drummed material and registration by producer rather than brand was the key to a nickel contract being introduced at all. Many references were made to the parallels between drummed nickel and the longstanding joke about "sardine can trading" on futures markets. As often happens, fears in advance of the event about sophisticated frauds being perpetrated with drums have, so far, proved exaggerated. The 6 tonnes warrant, incidentally, which is peculiar to nickel, would not have been thought of if the contract had been introduced in the fifties or sixties, as it is designed to allow warrants to be shipped three to a container. It should be borne in mind that the number of LME warehouses into which nickel can be delivered is more limited than for the other main metals, especially outside Europe.

STRUCTURAL CHANGE

Recent years have seen considerable changes in the nickel market structure, mostly on the supply side. A major feature has been political change in the former USSR. This has particularly affected the Norilsk operation in northern Russia, a mainstay for many years of that country's exports of nickel and an important component of its platinum exports. The ending of central state control of planning, price formation and even, to a degree, foreign exchange has affected all Russia's traditional export industries, but few as dramatically as nickel. Added to this, the opening up of the economy to outside interests has highlighted the environmentally unsatisfactory way in which the plant has been operating for many years. Neighbouring countries Finland and Norway have shown enlightened self-interest in helping Norilsk clean up its emissions. But even with generous injections of investment to this end, the programme cannot be completed without impacting production. Meantime, problems with infrastructure and cashflow make even the most simple operations of a producing works more difficult.

As always, the producers in Canada, Australia, New Caledonia and elsewhere in the capitalist system have had to ride the vagaries in availability of Russian, and at times Cuban, metal, on top of the normal fluctuations in industrial demand in tune with the health of the market. Fortunately, the intensity of nickel use in its major applications has not fluctuated much. As a mining-intensive metal, the geography of nickel supply is not normally subject to sudden or major changes. Also to be borne in mind is the fact that whereas some nickel has platinum metals as by-products, so some platinum mines produce by-product nickel. Such nickel has the most favourable cost structure of all and comes, notably, from South Africa and Russia.

PRICE FORMATION

It pays to have an understanding of the long-run history of the nickel market when studying attitudes and actions in the market today. In the early years of the industry in the nineteenth and early twentieth centuries, Sté Le Nickel was the only producer, based on its rich but inaccessible New Caledonian garnierite deposits. These are still an important part of world supply today. SLN's monopoly is now shared with some half-dozen private mines, though the latter do not themselves convert to ferro. World War I was a huge consumer of nickel in armaments, notably guns and ships, and put the Canadian industry on the map. After the war, the world surplus of nickel was so great that some mines – notably those in Canada – had to close entirely for two years until balance was restored. This experience scarred the industry, and influenced attitudes until after World War II. Meantime Ludwig Mond, anxious to form ICI as a giant merger of chemical concerns in the UK in the mid twenties, sold his Mond Nickel refining company to International Nickel of Canada, giving Inco a position of world dominance it retained until the explosion of laterite-based oxide output in the seventies. The name of the game was control, both of supplies and reference prices, and even at the time of the introduction of the LME contract in 1979, there was strong philosophical resistance to the idea from both producers and major consumers like the stainless steel works.

In fact, the producers had begun to lose their grip of the market as early as the late 1950s, as was reflected by the introduction of a regular "free market" quotation in Metal Bulletin from 1960. It was this trend, coupled with the determined and innovative attack on the supposed impossibility of defining a tradeable grade of nickel for a futures market, that vindicated the LME's action. Today the LME nickel price is accepted as the dominant reference price throughout the world, even in North America, where there is a natural reluctance to accept the relevance of a price that does not emanate from that continent.

CHAPTER 16

Tin

Tin is the second smallest-volume and second highest per-ton value metal traded on the London Metal Exchange. Compared with the other traditional base metals copper, lead and zinc, tin stands out with its 5 tonne warrant size. It also shares with copper the distinction of having another terminal market to consider. This is not the Kuala Lumpur Commodity Exchange, which has for some time been unimportant in global terms, and is now effectively dormant. It is the Kuala Lumpur Tin Market, which is a more active spot physical market; its quotations, or translations of them into other currencies, have some value as reference prices. But it still does no more than engage the LME's routine interest.

Nevertheless, tin has also produced the most notoriously educational upset in the history of the Exchange since the Secretan copper corner at the end of the nineteenth century. This was the collapse of the International Tin Agreement in 1985. With the benefit of hindsight, it is easy to see how, over the 30 years of intergovernmental Tin Agreements, the emphasis shifted from achieving price stability at levels reasonable for both producers and consumers to objectives consistent with a naked producer cartel. The reason the activities of the Agreement were not caught by anti-monopoly legislation anywhere in the world was that this was an intergovernmental agreement, not an intra-industry arrangement. However, it was this fact of support and involvement on the part of many governments, including some of the most powerful (even the USA was a member for one of the five-year agreements) that led some to continue trading with the Agreement's Buffer Stock manager at a time when, if his backing had been only private money, they would never have touched his business.

It took years to achieve a settlement, largely on moral grounds, of about a fifth of the Agreement's debts to its creditors at the time of its collapse. It is also widely believed, though it is not capable of objective proof, that the affair was responsible for the LME succumbing to Government pressure and abandoning its traditional structure as a principals' market in favour of joining the rest of the futures industry with the cleared market it has today. The debt fell heavily on the ring dealers, some of whom thereafter shut up shop, but nobody defaulted. It also affected several banks which had lent money against the collateral of the Buffer Stock of tin at the prices at which the Agreement itself had been supporting them. It is no wonder that the affair is still being spoken of ten years after the event, though whether the lessons to be learned have sunk in as effectively as they might is open to doubt. The producers involved, for instance, lost relatively little time in forming the Association of Tin Producing Countries, and through it continuing a programme of voluntary production restraint by its members, exactly on the model of the production controls that were a feature of the Tin Agreement. Despite this form of self-sacrifice, the producers have never enjoyed price levels since the collapse of the Agreement that compare with the levels built up under the last two Agreements.

Perhaps the remarkable thing about the collapse of the ITA is that it was itself the result of failing to note the main lesson of the Secretan affair

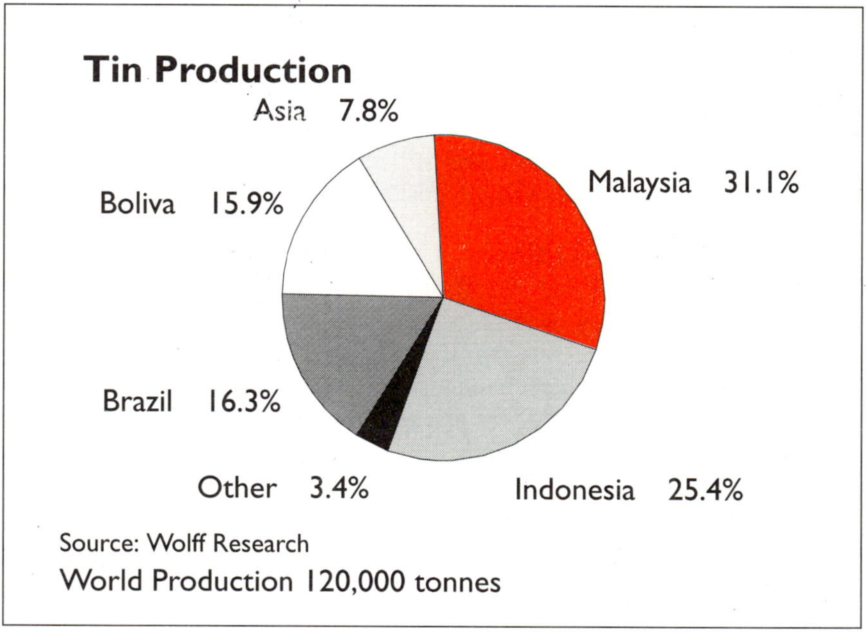

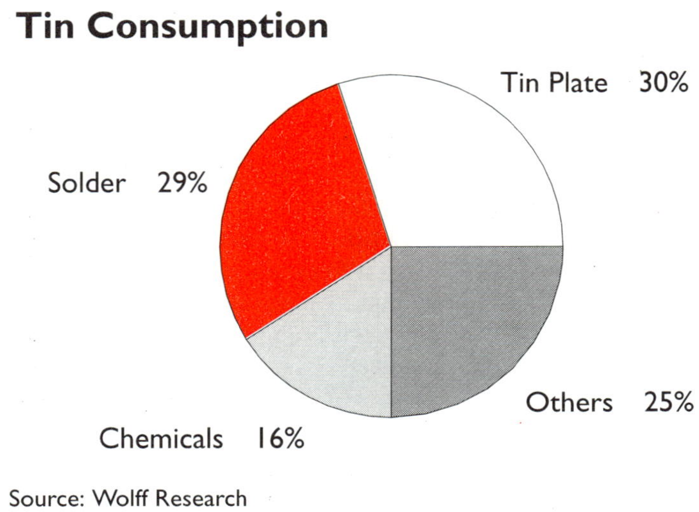

almost a hundred years earlier. This is that a cartel, to be effective, has to embrace all the sources of supply. In the case of the ITA, it was the failure to sign up China and Brazil as producers that sealed its fate. It should also not be forgotten that the USA has still not fully sold out its strategic stockpile of tin, accumulated in the postwar years.

CONSUMPTION PATTERNS

The backdrop of post-World War II macroeconomic forces against which the Tin Agreement story was played out contained the reasons for the Agreements being necessary in the first place. Despite a range of important and irreplaceable characteristics, tin has not been dealt with kindly by the modern world. Its principal use is in the manufacture of tinplate – a very thin sheet of steel coated with a microscopically-thin layer of tin. Nowadays, virtually the only application of this material is in food canning. Throughout the last fifty years, this application has been under competitive threat from all kinds of alternative forms of food preservation – freezing, cartons, plastic containers of various kinds and within the canning industry itself from aluminium. To hold on to markets in this low-cost sector, canners have gone for levels of efficiency and economy of use of their raw materials that would have been thought impossible fifty years ago. As the most expensive raw material by far, tin was the first to be economised. Continual reductions in the thickness of the tin layer were made possible by technological advances in the tinplate making process. Eventually, tin-free steel was developed, and has some role in the market, though its advance has been minimal since the crash in tin prices since the collapse of the ITA.

Even the steel component of the can has been downsized under the competition from lighter aluminium and double-reduced tinplate is almost half the thickness of the old standard. Even tin's special characteristic of solderability has been by-passed as the traditional three-piece can with soldered side seam and seamed-on ends has given way to the two-piece can, pioneered by aluminium, in which the body of the can is drawn from one circle of metal to which the lid is seamed on. Even this last has been lost to tin with the vogue for pull-top cans for many consumer products. Pull-tops need to be made from aluminium, as the tinplate cannot take the scoring necessary to make this idea work. At present the battle for market share between tinplate and aluminium is see-sawing back and forth,

Tin Commercial Stocks and LME Cash Prices

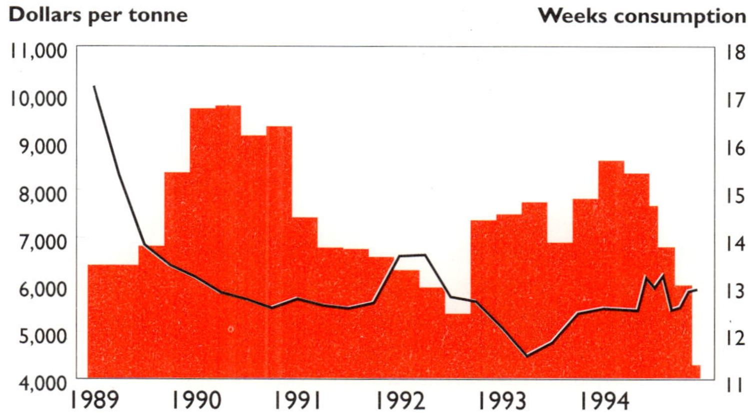

Source: Wolff Research

influenced by subtle non-market factors like environmental perceptions and the scale of the respective industries as employers in the various national economies, which affects the shape of political interference. All the time, competing forms of packaging and preserving are also nibbling at the market.

Of the other applications, solder is losing ground due to downsizing of components. Printed circuit boards may have large numbers of soldered connections on each one, but the unit consumption of solder is a fraction of what was used with manual assembly. The renewed drive to adopt aluminium radiators for cars is also hostile to solder use, whilst body solder, of which several pounds could be used in one bodyshell twenty years ago, is now rarely met, thanks to improved drawing quality of steel sheet and the development of resin-bonded fillers. A hopeful area of consumption is tin chemicals for applications such as pvc stabilisers, wood preservatives, pesticides and fungicides. The development of the float glass technique of producing plate glass has also been a boon for tin, though most of the demand is limited to starting up new works.

PRODUCTION

Tin production is not particularly demanding technologically. Mining is almost all by open pit with a high reliance on water. For large deposits, this takes the form of a pond in which a floating dredge operates. These massive items of equipment not only have a bucket chain arm like a conventional harbour dredger, they also commonly have on board the spirals and tables by which the tin is gravity-separated from the gangue in the ore. The dredge moves through the deposit by mining ground ahead of it and dumping its tailings behind it. Over time, a dredge may travel several kilometres in this way. Because of the immense cost of dismantling and re-erecting a dredge when it has worked out its deposit, it even pays for it to "swim" through kilometres of barren land to reach a fresh deposit. Some dredges have even worked in open sea, where the sea is sufficiently shallow and the seabed sufficiently rich in tin. Because most viable large tin deposits are alluvial and the tin contents are very low, the economies of scale of a dredge are needed. Smaller, slightly higher-grade deposits occurring in pockets in harder rock are mined hydraulically. High pressure hoses wash the tin-bearing earth into a sump at the bottom of the pit, from where the slurry is pumped up over separating tables and the tin concentrate is recovered. There is still a small amount of conventional underground vein mining for tin in hard rock in

Bolivia, but elsewhere in the world it is all over. This applies even to Cornwall with its thousands of years old traditions of tin mining dating back to Phoenician times.

Most mine production now comes from South East Asia, Latin America (notably Brazil) and China, one of the world's most traditional sources of tin. This geographical pattern is nowadays reflected in the disposition of the main tin smelting and refining centres, which are close to the sources of concentrates. Almost all the traditional European custom smelting industry, originally based on colonial sources of concentrates, has now closed down, whilst North America has never really had a normally viable tin smelting industry. There is, however, scope for small-scale secondary production in developed economies, based mainly on the harvest of an earlier industrial age which used tin alloys for bearings and wiped joints on lead pipes.

PRICE FACTORS

The prices emanating from the Malaysian market have the advantage of being formed earlier in the day than those in London or on the small inter-merchant market in New York. They are, therefore, always closely studied. But it is the LME with its much greater liquidity and openness to world influences which is the mainspring of tin price formation. Physical trade in Germany and the USA traditionally follows the KLTM prices, but otherwise the LME is dominant internationally.

CHAPTER 17

Zinc

Historically, zinc is an "old" metal like copper, lead or tin. However, it was not until the nineteenth century that it began to have any commercial life of its own as a metal. Prior to that, it was known mainly as the element that made copper into brass. Today, brass is still a meaningful outlet for zinc, but as an alloy with a high percentage of supply from recycling, the need for primary zinc to top up the recirculating volume of brass is relatively small. It was the use of zinc to protect steel by galvanizing that gave the metal its identity as an industrial base metal, and that is still the biggest use for zinc today. Along the way some other applications have been developed and appeared to promise much, but in the event it is still galvanizing that assures the future for high-volume consumption of zinc into the twenty-first century.

The other two main uses of zinc as a metal are sheet and die castings. Zinc sheet is a suitable metal for relatively low-cost weatherproofing of buildings as roofing sheet, gutters, downspouts and flashings. However, it has only ever become established in Continental Europe. Even in its heyday at the beginning of this century, zinc roofing did not seriously penetrate markets as nearby as the UK, where lead remained the preferred roofing material, or Sweden, with its tradition of copper roofing. In the USA, the use of terne plate (a steel sheet coated with a tin-lead alloy) was adopted from late 18th century England and remains the accepted method of roofing for historic buildings. Despite improvements in the zinc-base alloy now used in roofing, it appears to be in slow decline under competition from alternative roof-cladding methods; however, a glance at the skyline of cities like Paris and Brussels reveals acres of zinc roofing that requires periodic refurbishment.

The other application, which grew explosively in the sixties and into the seventies, is zinc pressure die casting. This was made possible in the late twenties by the development of electrolytic zinc of 99.95% purity.

Only by starting with metal this pure could the alloy maker be assured of eliminating the traces of lead that caused early attempts at die casting zinc to decay by intergranular corrosion. When some 3% of aluminium is added to the 99.99% special high grade zinc that is now always used for die castings, this produces an alloy capable of amazingly fine and accurate castings with good surface finish, strength, stiffness and reasonable properties as a bearing. All this is done at very high rates of production, thanks to the narrow freezing temperature range of zinc, and with very modest wear rates on the dies. As a result, zinc die castings are the standard method of producing intricate shapes like motor car carburettors, washing machine pumps and a thousand and one similar components in the automotive, domestic goods and electronics fields. In less functional applications, which were the basis of a lot of the growth in use of die castings in the sixties, zinc is now under remorseless threat from various compositions of moulding plastics. However, it can hold its own where highest accuracy, dimensional stability, stiffness and reasonable heat resistance are the criteria.

One of the effects of zinc's high dependence on galvanizing for its industrial consumption is the metal's relatively low recycling rate

Zinc Production

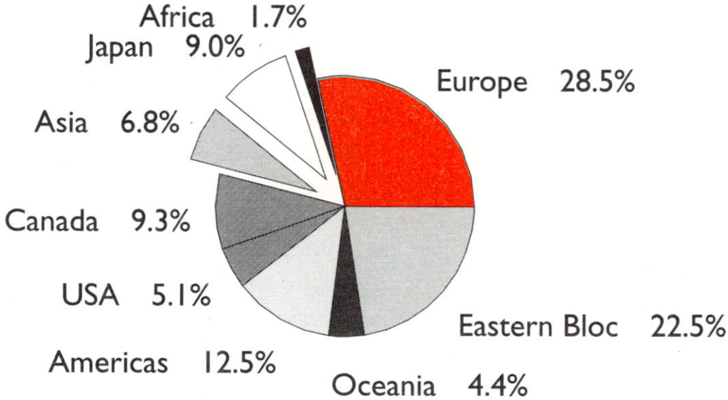

Source: Wolff Research

World production 7.4 million tonnes

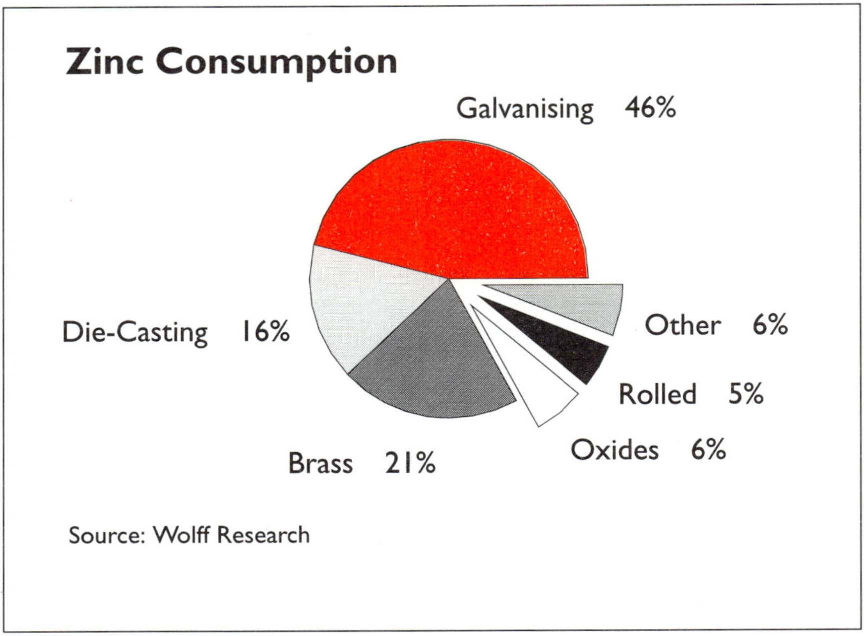

compared with lead or copper. The means by which zinc protects steel is its own sacrificial corrosion, so it can be said that after man has spent time and trouble reducing the zinc from its oxide, nature is busily converting it back to oxide! This does not mean, however, that every ton of zinc used in galvanizing is lost forever. In recent years, considerable efforts have been made to recover the zinc oxide flue dust that collects at arc and blast furnaces smelting scrap steel, including a good proportion of old galvanized iron. This flue dust then becomes feed for zinc producers. Other sources of zinc scrap are old die castings and zinc-containing residues from galvanizing pots and zinc melting furnaces.

In tune with its relative newness among the "old" metals, zinc was the last of these to be added to the LME trading range, in 1915; in those days it was known as spelter, a term familiar to those who haunt antique shops.

PRODUCTION

As is mentioned in the lead chapter, zinc is often mined in conjunction with that metal, and some output is sold as bulk concentrates with useful contents of both metals. In this form the concentrate can be smelted only by a works using the Imperial Smelting Process, the only surviving pyrometallurgical process, now that the horizontal retort, vertical retort and New Jersey processes have all passed into history. Like those processes, the ISP naturally produces a gob quality of zinc with 98% content. This quality is entirely acceptable for general galvanizing and moreover can be produced from a mix of primary and secondary feed material. However, by an extension of the process, high grade (99.95%) and even special high grade (99.99%) can be produced. In general, however, most zinc production today is produced electrolytically and is thereby hg (99.5%) or shg (99.99+%) quality, produced from clean concentrates of about 55% Zn. The LME contract does not specify that zinc delivered to the market shall be electrolytic, but with a purity specification of 99.995% it is not surprising all the registered brands are electro.

Both ISP and electrolytic processes start by roasting the zinc sulphide concentrate to produce oxide. For the ISP, this needs to be in the traditional clinker form and is then charged to the blast furnace. For electrolysis it is crushed or recovered from a flash roaster in more finely divided form and leached in sulphuric acid to form an electrolyte.

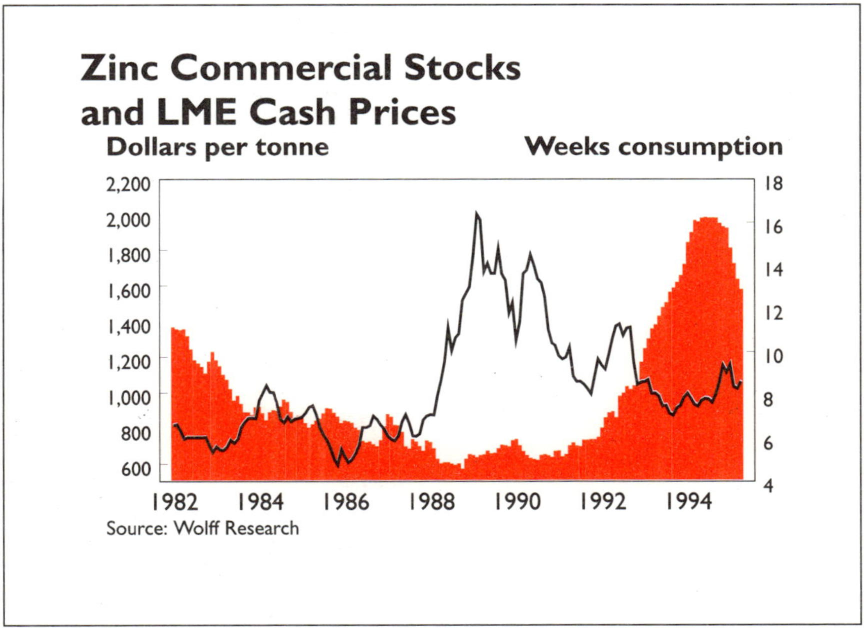

(Although zinc is described as electrolytic, it is strictly speaking electrowon). The sulphuric acid is conveniently largely produced from the off-gases from the roaster. The electrolyte is continuously circulated and refreshed in zinc content, while the electrowon zinc is deposited on inert cathodes (usually aluminium). Cathode zinc is not normally sold as such but remelted into 25-55 kg ingots or slabs.

PRICE FORMATION

Since the LME price stopped being controversial in itself by standardising on dollar pricing and shg quality several years ago, its influence in world markets has become almost total. The USA, with its strong commitment to the concept of producer prices, still operates on this basis for a good deal of its domestic business, but these prices effectively deviate only a little and only briefly from the LME price equivalent. In some ways, this is a remarkable situation when it is remembered that zinc is the only metal to have had the experience of an effective producer price displacing an established LME-related pricing structure. This happened in 1964 and lasted for some twenty years. The only sector of zinc trading outside the USA that never left an LME basis was scrap trade, and it was the leverage this exerted on brass semifabricators, together with the decline in relative importance of die casting, that started to undermine the influence of the producer price. Eventually, however, internal dissension between vertically-integrated and custom smelters proved the last straw. Perhaps this experience acted like an inoculation on the zinc industry, but it is certainly now “cured” of its enthusiasm for producer pricing in the traditionally LME-dominated markets outside the USA.

CHAPTER 18

Aluminium alloy

Aluminium alloy is in a number of respects an odd man out compared with the other six metals traded on the LME. To start with, it is the only contract commodity that is not a chemical element in its commercially pure form. It is also the only commodity whose market is a subset of what is normally considered another metal's domain – namely aluminium.

The aluminium alloy contract is based on three specifications which are broadly comparable, and all of which relate to aluminium used in castings. It is a peculiarity of aluminium that there is no secondary refining industry, as there is in copper and lead. This is why the alloys for castings are a discrete market from the applications for primary aluminium and the primary-based alloys. These latter applications are almost all in wrought form (sheet, extrusions, wire etc).

Thus an atom of aluminium, when first produced as primary will be fabricated either in the pure form as foil or electrical conductor, or as a wrought alloy as sheet or extrusion. When that wrought product is scrapped at the end of its life, the metal enters the scrap/casting alloys circuit and is re-formulated to casting alloy specification. Apart from beverage cans, very little wrought aluminium recycles back to wrought. Cast aluminium, on the other hand, can cycle round and round through the casting alloy ingot form indefinitely.

All this is in complete contrast with copper. In that metal, the existence of a highly-developed secondary refining industry keeps the markets in primary metal, scrap and secondary ingots quite closely interwoven. Aluminium has no secondary refining activity, so primary metal, scrap and secondary ingots have more scope to diverge from one another in prices and sentiment.

In both copper and aluminium, casting alloys are predominantly made from secondary sources, though a little production is based mainly or entirely on primary metal. This almost complete dependence on scrap as its raw material is another unique feature of aluminium alloy. Of the other LME metals, lead comes nearest with about half world output deriving from secondary sources. But then lead has an active secondary refining industry and a mine production.

Despite the critical role of aluminium castings in fields like transport, domestic goods and general engineering, the supply-demand statistics for aluminium alloy are notably less precise than for the other LME metals, mainly due to the scrap origin of most of the tonnage. Typical components made as aluminium castings are motor car cylinder heads, cylinder blocks, gearbox casings, brake cylinders and intake manifolds. Similar items are produced in much smaller volume (and therefore probably as sand castings) for marine and rail transport. In white goods, pumps, chassis, gearboxes and timer housings are usually die castings. Countless components in general engineering, healthcare, street furniture and agricultural machinery are made as aluminium castings.

THE SUPPLY PICTURE

Worldwide, aluminium alloy is an approximately 4m tonnes a year industry, making it roughly equal to lead on a volume basis, but some

Aluminium Alloy Consumption

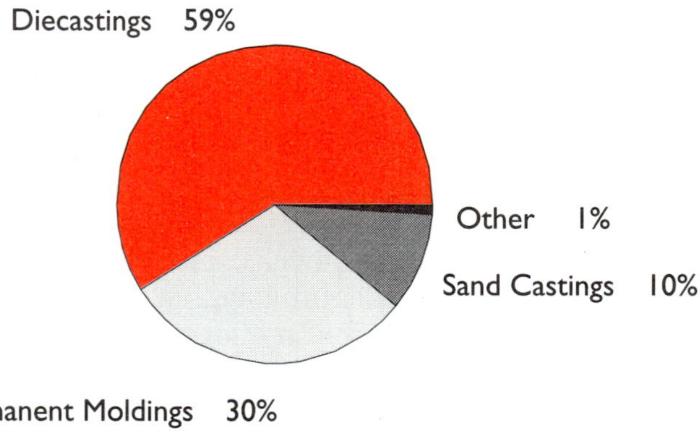

Source: Wolff Research

Annual market 4 million tonnes

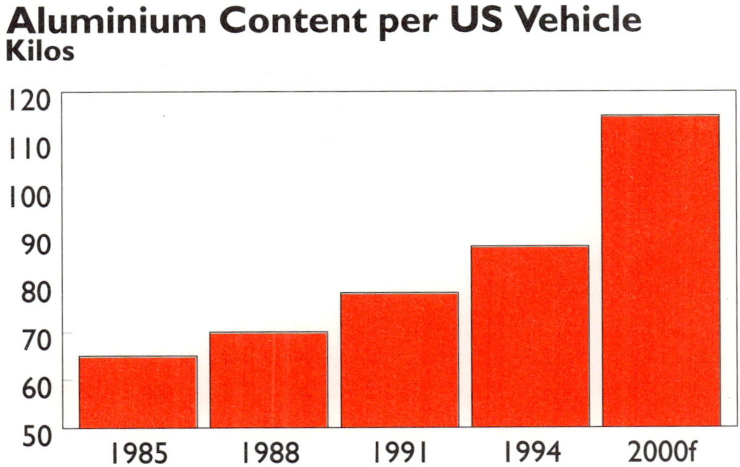

Source: Wolff Research

f forecast

three times bigger in terms of value. Consumption is dominated by the traditional industrialised areas – North America, Europe and Japan/South East Asia. However, production is also largely concentrated in those areas because of its scrap basis. As with several other metals, Russia is an uncertain element in the supply picture, due to wide swings in its internal supply-demand balance of both scrap and alloy.

Structurally, aluminium alloy has one of the lowest thresholds for capital cost of entry to the industry, this despite the growing burdens placed on scrap suppliers and ingot producers in the name of environmental protection. Ingot making companies rarely exceed 100,000 tonnes a year capacity. They employ both oil-fired furnaces (usually for small to medium-sized melts and flexibility) and electric furnaces (for longer runs of a standard alloy and reduced unit costs).

Aluminium alloy can be considered to have given the LME more difficulty in selecting a representative trading grade than any other metal except, perhaps, nickel. This was due to the large number of official specifications in existence. This in turn was due partly to the three basically self-contained regional markets in scrap and alloy in North America, Europe and South East Asia. This in turn led to different specification series prevailing in these regions. Partly, too, it was due to the fairly anarchic situation within these markets due to a proliferation of individual specifications traded between the ingot makers and foundries. This situation evolved because of the sometimes highly-leveraged cost implications of fairly small changes in content of certain impurities, tempting buyer and seller to agree to shade part of a standard specification in their bilateral trade. All this means that more than usual care is needed in taking up physical metal from LME stocks.

This in turn is not so much from the risk that the material will fail to conform to the LME specification for physical condition and chemical specification, as these aspects need to be tightly controlled by the warehouse and backed by the supplier's guarantees traceable through melt numbers. The risk is rather that metal exactly to specification may turn out to have a varying appeal to different consumers, depending on which way they may be in the habit of shading their local specification. This in turn, of course, presupposes that one is not trying to sell metal to the North American specification within the contract to a consumer in Japan, for example.

Clearly, the LME has succeeded in identifying a commodity grade of aluminium alloy lying somewhere in the middle of the spectrum of performance characteristics demanded by the foundries and end-users. As such, it is an entirely viable hedging medium for most of that spectrum of alloys and service applications. But it does mean that not all foundries would welcome LME metal as physical. It would be too poorly specified for high duty applications in aircraft or critical road or rail transport parts, and too expensive for a producer of cheap, low duty components. Consideration must also be given to the method of casting being used. The alloys in the LME contract are typically suitable for pressure die casting applications, so are generally also suitable for sand casting, if the price is right. For gravity and low- pressure die casting, however, the founder may prefer a somewhat tighter specification.

As this book goes to print there have been rumblings that some of the metal placed on warrant at the beginning of the contract's life does not meet the specification (which was slightly tweaked in the early stages). This has lead to calls for the whole specification question to be reviewed. After the efforts that went into selecting the US, German and Japanese specifications that are now deliverable, a wholesale structural revision is probably unlikely. Meantime, warrant-swapping, a familiar process in the older contracts, is now increasingly being used in aluminium alloy too.

SPECIAL WAREHOUSING RULES

Another of the aluminium alloy contract's idiosyncracies is its warrant size of 20 tonnes – a container-load – rather than the 25 tonnes of the other high-volume metals. And last but by no means least of the contract's unique features is the effective constraint that metal, once it has been taken off warrant, cannot be put back. This requirement is to support the principle that precautions must be taken with a material which is capable of suffering deterioration and oxidation of its surface during relatively prolonged storage. The conditions imposed on warehouses concerning protection from other stored materials and dust, as well as humidity control, are quite stringent, and regularly policed by the LME.

Historically, the fine points of aluminium alloy specifications and users' individual preferences have been part of the changing scene as the market moves from bullish to bearish and back again. In a tight market, users have been strangely less pernickety about the material they will accept than they were when supplies were plentiful. It has been a benefit of the LME's pioneering efforts in forging a commodity specification out of this kaleidoscope of ideas and prejudices that both producers and consumers of alloy have become somewhat more scientific in their use of the specifications as written. But it is still the case that market operators accustomed to delivering other LME metals as physical need to recognise that they are in a subtly different world when it comes to aluminium alloy.

When it comes to commercial practice in the marketplace, it has to be recognised that the LME contract, with its twice-daily fluctuations, had to gain acceptance in an industry with a deeply-entrenched culture of apparently more stable producer prices. In Europe, alloy makers and their customers the foundries were wedded to the concept of monthly prices, even though ingot makers never succeeded in forcing their scrap suppliers to adhere to the same structure. In the USA, the market was more accustomed to producers setting "price ruling" quotations for indefinite but sometimes extended periods. Both systems were prone to break down under conditions of exceptional market stress and rapid changes in either scrap purchase prices or prices obtainable for alloy in major export markets. Obviously, an appreciation of the scope for hedging aluminium alloy price exposure has had to wait on an increasing acceptance of LME prices as a valid yardstick for physical transactions, and this process is still some way from complete, not least in the LME's home market in the UK. However, a recent decision by the ingot makers' trade association to discontinue trying to form reference prices for use by its members and their customers reflects some continuing movement which could lead eventually to wider acceptance of LME prices.

In the contract's relatively short life it has progressed from initial industry hostility (to the point where many ingot makers would deny any interest in the levels being quoted on the LME) to the situation today, when no ingot maker in the world would deny the information value of the LME quotations to his physical business. This level of acceptance has, however, still to be reached among all the founders, and among end-users is largely confined to the largest automobile and white goods manufacturers. Meantime, the LME is sustaining the momentum by introducing traded options on aluminium alloy futures from January 1 1996.

As with primary aluminium, there is no doubt that in aluminium alloy, the LME has hitched its wagon to a star with generally better growth prospects than are typical for the older base metals.

CHAPTER 19

Historical background

History buffs will be interested to learn that the LME's first purpose-built home in Whittington Avenue, leading to Leadenhall Market, was very close to the site of the Roman forum in London. However, it would be a little tedious to trace the whole evolution from that marketplace through to the present day, especially as there is no record of the Romans dealing in futures, forwards or even arrivals. There are, however, grounds for believing that they knew about options.

For the purposes of this book, it is sufficient to pick up the story in the eighteenth century, when indeed trading in arrivals (the cargoes of vessels due to arrive) began to be a regular part of commerce. By the early nineteenth century, metal merchants regularly met one another in the Jerusalem Coffee House off Cornhill. Later, the walks of the Royal Exchange were a gathering point for all concerned with international business, and metal merchants would seek each other out among the crowds of financiers, ship charterers and other commodity merchants. In a bid for a select environment, they then migrated to the Lombard Exchange and Newsroom, where, however, they were seriously cramped. After a brief return to the bustle of the Royal Exchange, it was decided that metals deserved their own terminal market. On a historic day in 1876, a group of metal merchants, including Rudolf Wolff, decided to form the London Metal Exchange Company and it was established with a capitalisation of £1500. This started business in an upstairs room in Lombard Court, where it remained for some five years, trading copper, tin, pig iron and – irregularly – lead and zinc.

Compared with, for example, the Chicago Board of Trade or the Liverpool Cotton Exchange, the Metal Exchange was a relative latecomer to the business of establishing a formal terminal market. (It must be remembered that until very recent years the LME was, not always happily, referred to by other commodity futures markets as a "sophisticated forwards market, not a futures market"). By 1877, telegraphic advice of the day of arrival of a steamship was becoming the norm, and the vagaries of wind and weather that bedevilled the prediction of arrival dates for sailing ships were fast becoming history. Thus it was that, from its inception, the Exchange had the ability to trade in days rather than months.

In this way, it also established another dominant characteristic – namely a desire to match its facilities to the real-world needs of its clients in the metal producing, trading and consuming industry. This was reflected from the outset by trading in relatively narrowly-defined, but wholly representative grades of metal. These were Chile bars for copper and Straits for tin. Reflecting the importance to all commerce of prompt information, the Board of the new company ensured it was supplied with a telegraphic news service, and a telephone was installed in 1880.

Perhaps most importantly, the new room afforded scope for true open outcry trading, though originally this was due to nothing more than operating practice. Even the Ring, now regarded as so focal a point in the Exchange's life, developed organically and was produced ad hoc each day by chalking a circle on the floor. Equally, it was consensus rather than imposed regulation that produced the practice which distinguishes the LME from most other futures markets, namely the concentration of trading into rings and sessions, rather than an all-day affair.

Another driving factor behind the formation of the new company was the desire of members to see the transparency of open outcry translated into equally true and objective published prices. Taken very much for granted today, this aspect was to prove both controversial and influential in the early years of the Exchange. Rapid growth in the membership of the Exchange to some 300 members completely outweighed the original 15 Board members of the company, and the stirrings of democratic pressure began to be felt. In those earliest days, reports of closing prices and market activity were taken by the Board as their responsibility and privilege, but members soon began to express concern that this was not being done fairly.

Matters really came to a head when the Board decided, in 1880, to ban Ring trading on the twisted argument that open outcry trading led to manipulation of prices! After barely a year of a return to trading in huddled groups, members demanded, and succeeded in obtaining, the right to elect a Committee of subscribers, and the first such was elected in March 1881. For the next 106 years the business of the Exchange was run under this unusual dual form of governance. The Board remained responsible for administration of the rooms of the Exchange and for the financial health of the company, now called the Metal Market and Exchange Co. It was also jointly responsible with the annually-elected Committee of subscribers for decisions on membership. All other management of Ring trading through the progressively-evolved Rules of the Exchange and standard contracts was the responsibility of the Committee.

In 1882, the now quite substantial business of the Exchange was moved to its first purpose-built home in Whittington Avenue at the entrance to the Leadenhall meat and poultry market. For the next 98 years, Exchange members rubbed shoulders in the street and in the local pubs with market stallholders and porters, while conducting their internationally-significant and sophisticated business inside the dignified portals of the Exchange. Any market member in danger of forgetting that the Exchange was subject to the implacable law of supply and demand had only to study the movement of turkey prices in Leadenhall market in the last few days before Christmas – steeply rising in a year of shortage, falling in a year of surplus.

MONKEY BUSINESS

Unfortunately, however, there were those who thought that they could, if only temporarily, put themselves above that law and profit greatly from doing so. Thus it was that when the Exchange was barely ten years old, it found itself confronted with what has gone down in history as the Secretan Syndicate. The roots of this audacious plot, for that is what it was, lay in the fact that the original selection of Chile bars as the standard trading grade had been overtaken by rapid changes in world copper production. In particular, the US domestic industry was opening up, and the importance of Chile bars in international trade was declining steeply. In such circumstances, the market became heavily oversupplied, at the same time that the LME became thin and dangerously volatile.

Copper from sources other than Chile may not have been deliverable on the LME, but it was certainly saleable in the physical market on the same price basis as Chile bars. Inevitably, LME prices fell, until by 1887 the price of £38 7s 6d per ton was not much more than half what it was when the decade started. At this point, Pierre Secretan, manager of the Sté des Métaux in Paris, worked out that the narrow basis of the copper that was deliverable on the Exchange would make it easy to corner the market. Accordingly, he began buying on the Exchange in late 1887 and was soon able to start a bull movement in prices. In this he was supported by consumers who knew nothing of the corner, but observed prices rising and came into the market themselves. By the year-end, prices had risen to £85 a ton.

The new producers were naturally only too delighted to sell into this more remunerative market. And the aspect of the supply-demand balance that people can still overlook to this day – secondary refined metal – began to be attracted in ever-increasing tonnages. This meant the funds Secretan had to commit to building his stock of metal kept rising too. With backers of the substance of the Comptoir de l'Escompte, Rothschilds and Baring Brothers, he was able to dig himself well in to the tune of £2.5m, a huge amount at that date. He then began borrowing to extend his buying operation and by early 1889 the syndicate had borrowed £6m, yet the LME price was no higher than £75 a ton, £10 a ton less than a year earlier. Instead of completing its corner, the syndicate had to give up buying spot metal in the spring of 1889, at which time the secretary of the Comptoir de l'Escompte committed suicide. The syndicate then failed, copper prices fell back to £35 a ton, just below where they were when the exercise started, and the Banque de France moved in to bail out the Comptoir.

Looking across the years, the parallels with the failure of the International Tin Agreement almost a century later are frighteningly exact – except for the matter of national bank intervention. (Indeed, Secretan had also attempted some operations in tin as well). However, Secretan was not an isolated example of an attempt to flout the law of supply and demand. In the intervening years, there were quite a few others.

FRAMING CONTRACTS

Unwinding the problem of the Secretan stocks took the Banque de France a number of years, and meantime the LME had had to move positively to eliminate the problem of the excessive narrowness of its trading grade. In August 1888, when the Secretan squeeze was at its height, the Exchange introduced a new standard contract for Good Merchantable Brands. This not only broadened the geographical base of what was deliverable on the Exchange, it more importantly introduced the concept of registration of brands by the Exchange (specifically the Committee) which has been an identifying element in the Exchange's subsequent history as a trade-oriented market. However, the brands registrable included not only refined copper but also rough copper (an intermediate grade assaying around 96% Cu, with 93% set as the minimum). Rough copper remained deliverable – at a fixed discount – as late as the late thirties. Coupled with the seller's options on brand and warehouse location, it was not a popular form of physical and tended to lie in warehouse for extended periods. On the other hand, in the climate of Secretan – and of many other later occasions – the knowledge that there was a hardcore of LME stocks that would be very slow to move out into the market was comforting.

A third major innovation in 1888 was the standardisation of the forward trading period to three months. It is traditionally said that this was inspired by the shipping time for Chile bars to come to the UK, but it is noticeable that it was only formalised just as Chile bars were being removed from the standard contract. Delivery points were also formalised at the same time as London, Liverpool, Swansea and Birkenhead. Thanks to the pace of technological advance in the last decade of the nineteenth century, the standard purity of copper traded in the physical market kept rising – first to a norm of 99.75% for "best selected" fire refined and then to 99.9% as electrolytic refining began to be adopted.

This was reflected by yet another contract change in 1898 to a "Standard" contract based on fire refined of just over 99% purity, but with seller's option to deliver electro at a fixed premium or rough at a fixed discount. Originally introduced alongside the GMB contract, the new contract progressively ousted GMB, which was discontinued in 1902. Thanks, mainly, to a long period of relative stagnation in industry technology, this structure survived until the 1960's, though with technical adjustments to grades and fixed premiums and discounts along the way.

While all this was going on, tin, too, was benefiting from progressive clarification and refinement of its contractual trading basis. From the original Straits quality description, the market moved in 1891 to a Good Merchantable Quality contract covering both Straits and Australian tin (which is why it was GMQ, not GMB). This, too, proved dangerously narrowly-based and was supplemented in 1897 by a Mixed Tin contract which accepted not only Straits and Australian tin, but also the Dutch East Indies brands Banka and Billiton. Even now, the market was still susceptible to being squeezed, and it is tin that we have to thank for the fact that the LME rules include powers for the Exchange to take steps to avert a corner. These were introduced in 1899 after all other possibilities for broadening the base of the LME contract sufficiently to make a corner improbable had been canvassed and rejected.

A further attempt to follow the lead of copper and introduce a standard tin contract was made just before World War I against a background of continuing volatility in LME – and Far Eastern – prices. Once again, it was introduced alongside the incumbent GMQ contract, and, as with copper, gave the option of delivering an inferior grade at a fixed discount. Straits and Australian tin, assaying 99.75%, were now called Class A and the new Class B covered English and other tins of a minimum 99% quality. The often unpopular new contract had hardly settled in when World War I started and the Exchange voluntarily closed down for three months (the then-expected duration of the war). It reopened, but metal trading was far from normal, and in 1916 the Government took over all distribution and allocation of metals. The period was a tense one for the Exchange, as many senior members were of German nationality, family or extraction.

Lead and zinc were still not really "accepted" LME metals, though the unofficial trading in lead in a separate ring benefited from the introduction in 1903 of a Good Soft Pig Lead contract which gave some shape to quality requirements. However, it was not as precise as the copper and tin contracts, did not rely on brands and was not for in-warehouse trade, being still essentially an arrivals contract with delivery in the Port of London Authority area. As a result, settlement was for a delivery month, with a further half-month's period of grace. This ethos was to remain with lead and zinc until well after World War II. The LME had its first zinc contract – then called Virgin Spelter – in 1915, modelled on the Pig Lead contract but with delivery points in Liverpool and Swansea as well as London.

One of the factors inhibiting earlier trading of zinc on the LME was the formation in 1908 of the European Spelter Convention, a cartel comprising the German, Belgian, French and British producers, who agreed to operate production restrictions. In the same way, a Lead Smelters' Association sought to control the European lead market by centralising marketing of German, Spanish and Belgian metal through Metallgesellschaft, which also entered, effectively, a non-competition agreement with the US producer Asarco. Unlike zinc, this cartel was still fairly successful in its objectives as World War I broke out.

AFTER THE WAR WAS OVER

The LME resumed normal trading as soon as the war ended – in December 1918 – and was soon caught up in the postwar rebuilding boom that preceded the Great Depression of 1929. During this period the Exchange had its first taste of the impact of currency instability on commodity prices through the use of commodities as currency hedge vehicles. One effect of this mood was to bring lead and zinc "into the family" and they were traded in the main ring alongside copper and tin from 1920. By mid-decade, the shadow of the Depression was being cast before it, metal prices were no longer buoyant and the cartels again reared their heads.

Best-remembered of these is the US-inspired Copper Exporters Inc, which traded on the dominant position of the USA as a world supplier. Not only was US native production the biggest national output in the world, US money was behind most of the Latin American production; and the fledgling African production (from Zaire only at that time) was largely refined in the USA. Together with the major Spanish and German outputs, the cartel represented 95% of world primary copper output. The operation brought into the open the endemic hatred of US miners for futures markets and their prices (Comex was then still ten years in the future). This influenced their commercial policies until the 1970's. Copper Exporters' object was to sell direct to consumers at its own (more or less) fixed prices, and to by-pass the LME. So obsessed was CEI with this aspect that it would supply consumers only with their day-to-day requirements, so that they should not accumulate stocks and re-sell them on the LME. Again, this spirit was still at work in the sixties, when US producers' agents would spend much time detecting how, and through whom, any parcels of their own brands of metal came to be in LME warehouses.

Having run down LME stocks and emasculated the exchange as a hedging medium, CEI began to apply the thumbscrews and kept moving prices ever higher until the music stopped abruptly with the Great Depression. Once again, the unlearned lesson of the Secretan Syndicate was repeated, and declining consumer demand was aggravated by a surge in scrap and secondary availability. Inevitably the cartel began to crumble; US custom smelters flagrantly undercut the cartel prices and LME prices openly reflected what was happening. To turn the knife in the wound, the Rhodesian Copperbelt (now Zambia) was opening up its rich deposits. High cost US domestic producers were the first to feel the inevitable collapse in prices and by 1932 US output was only a quarter of its 1929 level. At this point, with copper prices around 4c per lb, the US isolationist streak resulted in a 4c per lb import duty there, which at least protected the domestic market for the surviving domestic producers. It didn't, however, absorb those producers' unsold stocks, and these began to flood on to the world market.

TIN AND BUFFER POOLS

Once again, tin moved in parallel with copper – indeed, ahead of it – in forgetting the lesson of Secretan. As early as 1920, the Dutch East Indies (effectively today's Indonesia) and Malaya (mainland Malaysia) formed the Bandoeng Pool, so called after the Javanese town where it was conceived. This appeared to manage the next cycle in prices quite well, accumulating stocks and helping stabilise prices until 1923, when it began to run its stockpile down. However, in reality the Pool had given tin prices a floor which encouraged additional production. As this came on stream in 1929, it was just in time to meet the Depression and prices began to fall again. The newly-formed Tin Producers Association began to persuade its members to adopt production restriction on a voluntary basis, but not surprisingly this was not powerful enough.

In 1931, therefore, the International Tin Restriction Scheme was launched, run by an International Tin Committee. Once again, however, greed overtook prudence and production was so far restricted – by 66% – that prices indeed recovered and by the end of 1933, a year after the draconian restriction was introduced, were virtually double those of a year before. In 1934 the scheme was renewed for a further three years with the added twist that it would revive from the Bandoeng scheme the idea of a buffer pool and use it to stabilise the market at £230, the prevailing price. This would be done by buying tin when it was clearly below £230 and selling it when it was above.

The LME protested to the Colonial Office that the scheme was inimical to free markets, but it went ahead anyway with members contributing an initial 8,000 tons. In addition, production continued restricted to 40% of members' "standard" outputs. Although the Committee, through its pool manager, would operate on the Exchange, it publicly criticised it as "a playground for bulls and bears". Meantime LME stocks declined to unprecedentedly low levels, prices shot up and there was a wide backwardation. In 1935, production quotas were twice raised until they were nearly back to "standard" levels. Stocks recovered, the back narrowed and the scheme was renewed at the end of 1936. As the armaments race led a recovery in the world economy, consumption boomed, the buffer stock was run down and by March 1937 was exhausted. A new pool was set up in 1938, but was already exhausted again by the time war broke out in 1939.

LEAD-ZINC BETWEEN THE WARS

The vogue for some sort of cartel control of markets extended also to lead and zinc. Zinc was the first to move with the formation of what proved to be a short-lived and abortive Zinc Cartel (1928-29). This was followed in July 1931 with a more serious International Zinc Cartel to which all the non-US producers adhered. This, however, ran straight into the wave of protectionism that bedevilled 1930's trade, including UK Empire Preference tariffs, which presented the LME with the problem of different tariffs on the same metal, depending on origin. These lasted right up to the UK's entry to the EEC, and were dealt with by obliging LME ring dealers to charge clients an "Empire premium" on warrants of empire metal taken out of warehouse, in order to equalise them with the dutiable warrants of non-empire material. This prevented the duty question impacting on trading across the ring. In those days, of course, all warehouses were still in the UK.

Even lead, with its high secondary component, did not escape the fashion for cartels and the Lead Producers' Reporting Association was formed by non-US producers on the model of the zinc cartel in 1931. Here, too, the introduction of protectionist duties in 1932 caused the group to be disbanded in 1932. In 1938 a Lead Producers' Association was formed to operate both production restrictions and a buffer pool, but it had little chance or cause to do much before the outbreak of World War II.

There is one way in which the LME is itself a footnote to the economic history of the period, and this is in respect of the nature of the Empire protection duties on lead and zinc. Originally introduced as ad valorem duties, they were thus particularly inimical to normal hedging operations. After discussion between the Exchange and the Government on this point, they were changed to flat rate duties, a perfectly workable arrangement in those zero-inflationary times.

WORLD WAR II – THE LONG NIGHT.

When World War II broke out in September 1939, copper, lead and zinc trading ceased immediately and these metals were taken over by the Ministry of Supply. Tin trading continued as best it could until December 1941 when Japan entered the war and the fate of south-east Asian supplies became extremely uncertain. During the war a good deal of work was done on reviewing and revising contracts to be ready for the resumption of trading in peacetime. Nobody, however, predicted that it was to be 1949 before the first postwar trading (in tin) would take place. Even less could they predict that the last metal to resume trading – copper – would not do so until August 1953, over eight years after the end of the war in Europe. Part of the reason for this was functional, notably the postwar need for currency controls on sterling; but part has always been thought to be connected with a doctrinaire dislike of free markets by the postwar socialist government.

It was, of course, impossible to maintain fixed prices in the economic climate of the late forties, so the UK government took the available fluctuating US dollar prices as its reference for the allocation of metal from its stocks. In one sense, this was justifiable, as the USA was now an importer of base metals which had effectively become dollar commodities as a result. The exception to this was tin, which continued as a sterling commodity as Malaya was still in the sterling zone. The administration of metals control during this period was not particularly adroit, but its ending was nevertheless regretted by a generation of UK manufacturers who had known nothing of the norms of free markets and the ability to hedge for 14 years.

In addition to this, the UK Government ended the war with significant surplus stocks of tin which it needed to turn into cash. For this purpose, a reopened LME was a suitable vehicle and JD (Jimmy) Wolff was duly appointed the first Government broker. He was a natural choice, having been chairman of the Committee every year bar one from 1928 to 1950 and of the Board from 1946 to 1961. The position was more honorary than profitable, as the commissions earned on skilfully disposing of the Government metal were then shared out between all ring members!

RETURN TO ALL – TOO NORMAL

With the return of a Conservative government there was a fresh impetus for the resumption of normal trading and indeed this happened for lead on October 1 1952, zinc on January 2 1953 and copper on August 5 1953. First, however, a modus vivendi for the international clientele of the Exchange within a currency-control environment had to be worked out. This was successfully done with a modicum of trust and thus the minimum of bureaucracy under the so-called Bank of England M (for metals) scheme. Briefly, the Committee underwrote the good faith of the ring dealing firms, each of which was individually enrolled in the scheme and obliged to make returns monthly to the Bank of all its transactions that involved transfers, in or out, with non-sterling clients. Apart from the cachet it gave LME members, the scheme also involved regular meetings between the Bank and the Exchange which cemented personal relationships and created a climate of understanding that were to prove useful in the future.

In one way, that future looked remarkably like the past, in that moves to modify or frustrate the working of the law of supply and demand were soon back on the menu. As befitted the first metal to resume trading postwar and the one with the most continuous prewar history of market management schemes, it was tin which was first in the field. The big difference compared with prewar years was that the USA now had really stringent anti-monopoly legislation, and other countries were beginning to head down the same path. As a result, the structurally simple arrangements between companies of prewar had to be replaced by intergovernmental arrangements in order to put them above the reach of national legislation. Fortunately, there was a new body, the United Nations, with an agency called Unctad (United Nations Commission on Trade and Development) ideally constituted for the purpose.

Thus when the tin producers, once again led by the south-east Asian countries, wanted to get together, it was under the aegis of the United Nations that they did it. Several years of talks and procedural delays finally led, in 1956, to the start of the first of a series of five-year International Tin Agreements. With a heavy dose of déjà-vu, the ITA was managed by an International Tin Council (ITC) with a Buffer Stock and a Buffer Stock Manager. Like the prewar arrangements, the Agreement relied on both production restriction when necessary and a buffer stock to keep prices in a range that would be "fair to both producers and consumers". To this end, consuming as well as producing countries were embraced and each bloc was given an equal total of votes. The final twist of extra complexity compared with prewar was that the Buffer Stock Manager was given three specified price ranges which governed how he was to operate.

In the middle was a "normal" range of prices within which he was to take no action. Below this was a price range within which he was told he may buy to nudge prices back into the middle range. Above the middle range was the "may sell" range. If prices breached the bottom of the "may buy" range, he had to buy for all he was worth until either prices recovered or he ran out of money. If they breached the top of the "may sell" range, he was obliged to sell until he turned prices round or ran out of tin. It sounded very scientific. It is not possible here to detail all that happened through five five-year Agreements. Suffice it to say that, along the way, the Agreements encountered an empty Buffer Stock and prices soaring out of control, an empty kitty and an overstuffed Buffer Stock and a strange and inconclusive scandal relating to a buffer stock manager. What is important is to record what happened in the Sixth Agreement, beginning in 1981. (For those who want to study the matter in greater detail, a whole book has been written about it called "The Tin Men", also published by Metal Bulletin Books).

THE TIN CRISIS

With hindsight, it is easy to say that the postwar Agreements reached a high with the Fifth, to which the USA and Germany were consumer signatories. But it is also possible to say that the process of edging up the price ranges to give rather more favour to producers than consumers had already begun, not least because the Buffer Stock price ranges had been converted to Malaysian ringgit numbers from the sterling references with which the whole process started. As a currency linked to the dollar, the ringgit had been strengthening relative to sterling, effectively enhancing the Buffer Stock Manager's buying power on the LME. In short, with hindsight, the levels he was defending under the Sixth Agreement were too high. What is more, the buffer stock for that Agreement had never been properly funded with cash, as it had started operation without ever having been properly ratified by many members, who had accordingly made no contributions.

As a result, the buffer stock manager was forced increasingly to support the market by borrowing against his tin holdings, just like Secretan a century before. Because he represented an inter-governmental Agreement, banks continued to finance him, and brokers continued to accept his orders, believing his credit to be fireproof. On October 24 1985 they learned otherwise. When the buffer stock manager baldly announced that he could not pay his differences, the shock to the system for the LME and the banks was great enough. But it was not until a few days later that it sank in that the Governments behind the Agreement had no intention of honouring their obligations, and that the market had been left with a £900m hole in its accounts. Trading was suspended with tin falling precipitously to £8140. With the overhang of some 60,000 tonnes from the Buffer stock, the "real" price was about half this level.

As the grim postmortem into the debacle began, it became clear just what a house of cards the Buffer Stock manager had been obliged to create during the Sixth Agreement to meet his primary obligation to support the market according to the prescribed price ranges. Apart from the borrowings against his existing holdings that had funded much of the extension of the stock, he had also entered "special lends" and "special borrows" deals with brokers and banks which allowed him to keep the metal involved "off the balance sheet" as far as official buffer stock holdings were concerned. In the principals' market of the time, there was even less chance than at present that any one bank or broker could know how widely the buffer stock manager was repeating the transactions he was concluding with him. In fact he had eight principal channels, which was a surprise to everybody.

The LME called its members who were ITC creditors for security deposits and £52m was paid in – an earnest of the Exchange's commitment to seeking a solution. However, the reaction of the governments who were members of the Agreement was virtually the opposite. Irresolute, but clearly determined to disclaim any liability for the problem they sank into a mire of obfuscation. Concrete proposals from the creditors, thrashed out in late night meetings, appeared to progress, only to fail on a technicality or the total unreasonableness of one or a few governments. The LME contemplated the risky but clinically clean solution of resuming trading – the "free" market at the time was now around £6000 a tonne – but was dissuaded, notably by the UK authorities, clearly concerned that the ultimate downside was the destruction of the whole LME.

Finally on March 12 1986, the LME instituted a "Ring-out" – posting a figure of £6250 a tonne at which all open contracts had to be settled. Before it could do this, it had first to write a new rule empowering itself to do so by an extension of its existing powers for dealing with crises such as an impending corner. The new rule – Rule M – featured prominently in the subsequent court cases that were, perhaps, inevitable when a nettle of this magnitude is grasped. Several years were to pass before these cases were settled – all, it may be said, in the LME's favour. It was then possible to resume LME tin trading, which was done in 1989, and for men of goodwill to try yet again to act as honest brokers in the quest for a settlement that would put the affair to rest. Eventually such a settlement was thrashed out – still against the treacly prevarication of the various governments involved – and some £150m was paid in to what had originally been a £900m hole. If interest had been running on the original debt the settlement would have shown up as little better than 10p in the £; and that is without counting the legal costs of the various court cases.

A SEA CHANGE

All this was not the end of the affair for the LME. At the time of the tin collapse it was already coming under pressure from government to relinquish its cherished, traditional, but by now distinctly unusual principals' market structure. With just about every other commodity futures market in the world operating as a cleared market, and with the same structure normal for the financial futures markets that were then proliferating around the world, the LME's principals' market was a decidedly untidy feature in a landscape into which the Financial Services Act was being introduced. Even without the tin crisis, it is doubtful if the LME could have held out by reciting its argument that the principals' market demonstrably wasn't broke, so there was no need to fix it.

As it was, the tin crisis, through no fault of the LME's and in a syllogistic way, pulled the rug from under that argument and on May 29 1987 a cleared market was introduced. To make way for that, and to separate the ongoing business of the Exchange from the old Metal Market and Exchange Co, still the subject of litigation, a new company, London Metal Exchange Ltd was formed. Out of the same window as the principals' market went the dual structure of Board and Committee. Today the Exchange is ruled by a Board of prescribed composition, drawing from ring dealing members, the trade and consumer interests; administration is in the hands of a professional Secretariat under a chief executive.

Having followed that story through to its conclusion, we should now revert to our chronological sequence and return to the swinging sixties. In the light of manipulations recorded so far, it will come as no surprise to learn that tin was not unique in still trying to outwit the law of supply and demand.

IDEALISM ARROGANCE AND FUTILITY

Unhappy at this situation, certain of the major producers sought to revert to selling their copper at fixed prices. The first unilateral attempt by a producer to introduce a producer price for European business was made by International Nickel in 1954. It was abandoned after one year, as the producer had made considerable sacrifices on its fixed selling price for its (by-product) copper without having any impact on the market.

However, the experiment did not deter the powerful Rhodesian Selection Trust from introducing its own 'RST price' in 1955.

This stable 'price ruling' producer price based on the US model, carried with its general discount under the LME an obligation on the consumer to agree to price his semis at a strictly related price. This carried forward the split from the rest of the industry, which continued LME-based, into the semis sector. Not surprisingly, such strain could not last, and the experiment was abandoned in mid-1957. The record shows that the price was, on a month on month basis, very little less volatile than the LME quotation at the time.

One of RST's strong fellow-thinkers on prices was the Belgian SGM selling Union Minière copper from the Belgian Congo, now Zaire. It introduced its own producer price, and maintained this most assiduously for very many years, while taking the precaution of adjusting it a little every few days in order not to get too far out of line with the LME. The whole experiment led to one of the greatest exercises in futility ever seen in copper prices in the shape of the 'Metal Bulletin cif Europe wirebar average price'.

In a nutshell, this reflected the producers' belief that the ills of an LME-related European copper market were attributable to the lack of an instrument for European use comparable to the 'E & M J Metal and Mineral Markets' weighted averages used by the US industry). That journal is now known as "Metals Week").

They therefore invented one. For eight years, Metal Bulletin faithfully calculated, on behalf of the producers, a series of daily and weekly weighted averages based on reports of their sales by some eight to ten major producers selling into Europe. The reports were all adjusted to a wirebar basis by the producers prior to reporting, while the Metal Bulletin assiduously chased them up to prepare the average for printing in its twice-weekly editions. Despite all this effort, there is no record that a single ton of copper ever changed hands indexed on this average.

However, by late 1961, the producers were back on the attack with a much more intelligent scheme to stabilise the market through the LME, rather than by-passing it. Indeed, the scheme was much more dangerous for the Exchange, because it involved holding the market at an unchanged price by operating on it within very narrow limits.

The cash wirebar price used was £234 per ton, and after some early manoeuvring in the following six months the market was held unchanged at this level from the spring of 1962 until the end of 1963.

It says much for the LME's powers of survival that an institution to which movement, and the fear of movement, is meat and drink was not wiped out by such a long spell of predictability. In fact, LME houses survived on brokerage in other metals, and it is not surprising to recall that this was the first time that an LME aluminium contract was mooted, though it was not introduced until over a decade later.

THE BATTLE OF THE PRICES

The affair became celebrated as the battle of the prices and was closely studied at the time to see if it would serve as a model for a wider range of commodity stabilising schemes. The rival ideologies of free market and producer-controlled stability were extensively aired.

In the event the free market did break out around Christmas 1963 and by mid-January 1964 it was all over. The producers gave up controlling the market, but introduced their own producer price, essentially a joint pricing move by the two Zambian producers, which tried to be influenced more by the US producer price than the LME. This lasted a further two years before it was abandoned at the specific request of the consumers. After their taste of stability, they decided it was not as nice as they had imagined, and that the familiar volatility was certainly to be preferred to a second acquaintance with the strains and distortions of two reference prices side by side.

With hindsight, several lessons were drawn out of the experience. One of the most important was that the LME is a lot more durable an institution than anybody ever imagined. Certainly, its future was much more severely tested in 1963 than by the 14 years of suspension of its leading metal, copper, from 1939 to 1953 because of the war. Another was that price stability does not completely alter a metal's competitive position against others. Substitution of copper by aluminium in the electrical conductor field continued pretty much at the same rate throughout the sixties.

However, the most important lesson was the vital place in the copper market of the LME's daily prices. It highlighted the fact that the ups and downs of LME prices as a continuous sensitive indicator of the market's health were severely missed during the period of enforced stability. When the producers first intervened on the LME, its trend was a downward one.

In fact, however, the statistical balance of the market had turned, not least thanks to the production control by some producers, but this was not yet reflected in the statistics. With a normally vital LME, the market would have picked up the early signs of this swing without the help of statistics. As it was, a few LME operators, highly sensitive to straws in the wind in the market place, plucked up courage to test the producers' grip of the market, not by offering surplus metal, but by bidding above £234.

The producers' brokers met the bid, but the ice was broken. More operators decided to buy and the brokers soon ran out of ready metal. Meanwhile consumers too began to call for more metal from the producers. In less than a month from the first buying bid on the LME the producers announced that they had abandoned their stabilisation programme and were now quoting a producer price independent of the LME. That lasted even less time than the RST price, being abandoned by all in the spring of 1966 after it had indirectly been the cause of the high and unstable prices on the LME which had brought about its very abandonment.

In short, what turned out to be artificially low fixed prices starved the LME of offerings of stocks adequate to prevent sharp increases in its own price occurring when such marginal supplies were suddenly seen to be scarce. These very increases were the cause of the hasty upwards moves in the producers' own price which became increasingly out of step with the LME. Had the majority of the producers not chosen to fix their prices in quite such an arbitrary manner, the free market as expressed by the LME would have moved more sedately and in a less extreme manner in reflecting changes in supply and demand over the period.

AND A TALE FROM ZINC

A further example of producer pressure occurred in the zinc market in the early 1960s. At this time zinc was a very strong market on the LME. Trading had become more fluid with the cessation of certain artificial restrictions, and especially with the replacement in October 1960 of the previous ex-ship contract with daily settlement in line with other metals. Rotterdam was then shortly added to the list of approved delivery points for zinc.

Although there were occasional dips, LME prices continued very firm, to the point where (once again motivated by fear of a loss of business to substitutes) the Imperial Smelting Corp (now Britannia Zinc, a subsidiary of MIM) issued a statement in July 1964 claiming: "In view of the negligible tonnage of zinc stocks remaining in London Metal Exchange warehouses, the principal producers supplying the UK and Continental markets consider that the price quoted on the London Metal Exchange has become unrealistic and most harmful to the long term interests of the consumer industries and therefore also of producers. Imperial Smelting Corp have accordingly today decided to offer their customers zinc metal against all new sales and unpriced balances of existing contracts at a base price of £125 a ton. This price is effective today and will remain in force until further notice". The LME price at the time was £139.

ISC's move was followed by other producers and the situation thus became comparable with that obtaining in copper, but not altogether so. Unlike the virtual stagnation of the market which followed the fixed copper price of £234, the zinc market on the LME continued to trade actively and to record its moving, free market price. In fact the LME quickly fell below the producer price, which in turn was lowered from £125 to £110. A major factor in the continued parallel activity on the LME was the refusal of the scrap trade – whether concerned with zinc scrap and residues, or with the zinc component of brass scrap – to pay any attention to the producer price.

Nevertheless the producer zinc price, or pzp as it became known, had a better span of life as an effective reference price for business in Europe, and ultimately much of the world outside the USA, than many of the others recorded here. It was not finally abandoned – again by mutual consent between producer and consumers – until more than twenty years had passed.

SOME SINGLE-HANDED MANIPULATIONS

It may appear from this recital of major incidents in the Exchange's history that the attempts to manipulate the market came solely from groups of corporations or governments. However, postwar LME history also includes two famous individual manipulations.

Far and away the most conspicuous of these was the attempted silver corner in 1979 and 1980 usually associated with the name of Nelson Bunker Hunt, though other speculators in the Middle East and Brazil were also widely believed to be involved. Silver is not now traded on the LME, but at the time the LME silver contract was an important alternative silver futures vehicle to the dominant Comex contract. Bunker Hunt in fact used all available markets, including the London and other bullion markets.

The Bunker Hunts were no novices in silver, as this had long been a favourite vehicle of theirs for hedging some of their fabulous wealth. So it is the more surprising with hindsight that Nelson Bunker Hunt should have attempted an operation so similar to the Secretan copper corner. With such a large store of silver aboveground as coins and artefacts, it was inevitable that the high prices brought about by the Bunker Hunt squeeze should bring large tonnages of the metal "out of the woodwork" until the offerings overwhelmed even the huge resources of Bunker Hunt and his associates.

The other moral tale of the precariousness of attempting to manipulate free markets forms a sub-plot to the Tin Agreement story. In the second half of 1981 manoeuvrings began to be discerned in tin which clearly had the aim of bulling up a largely trendless and lacklustre market. Again, with the benefit of hindsight, we can unravel what at the time was a most mystifying, and often mysterious, episode. With producer members of the ITA suffering 40% production restriction, yet a serious statistical surplus which gave little hope of relief, it is easy to see how a large producer could be persuaded that a manipulation of the LME to drive up futures prices could be distinctly remunerative for its sales indexed on that quotation. There is little doubt that the initiative in framing the scheme came from one individual.

Even on an open outcry market like the LME, it took some days to pinpoint which broker was leading the buying operation, and some further weeks before the finger of suspicion could light with any certainty on the trade client responsible. Subsequently as the operation was extended through sophisticated technical manoeuvring it became clear that the aim was not a quick kill on the LME by a speculative client. Increasingly the view gained ground that the move was producer-inspired, with the trade house merely providing the technical expertise. Despite explicit denials, the suspicion grew that the Malaysian government-owned Malaysian Mining Corp was the real force behind the operation, in concert with the Indonesian producer. We had, however, to wait four and a half years before the Malaysians finally admitted officially that they had been the intended beneficiaries and the big losers.

As with operations by the Tin Agreement buffer stock manager, the operation attempted to wrong-foot the market by suddenly switching support from the forward position (the normal way to support a market because it can all be done on margin funds) to cash. This turned the market from a substantial contango to a large backwardation. Speculators and trade hedgers sensed the impending collapse of the support manoeuvre and took appropriate steps (aggressive and defensive respectively) by short-selling in the market. Three months later, as the end of February approached, a severe supply squeeze inevitably emerged. The backwardation opened out whilst remaining concentrated in the few days leading up to February 25 and 26.

The dilemma for the LME authorities was straightforward: were they confronted with an impending corner which the rules enjoined then to avert, or weren't they? They decided they probably were, so called for members' cards to find the nature of the exposed positions. In the light of this, they imposed a £120 a day backwardation limit. There was, of course, a lot of grumbling, but the balanced historical view is that this was a case where action was probably preferable to inaction.

The main point to be made about these attempted manipulations is that, despite disrupting trading conditions in the short term, they eventually failed. It is thus a tribute to the mechanism of the LME and of free prices that orderly trading was subsequently resumed.

MOVING HOUSE – AND SOME GOALPOSTS

If Rudolf Wolff could have returned to Earth in 1963, he would have felt quite at home on the Metal Exchange in Whittington Avenue. The building, furnishings and decor were unchanged from his day. All was stained mahogany and frosted or bevelled glass; the Secretary's office was up a narrow flight of stairs in the adjacent alley. In 1964, however, a structural alteration to the building was the cue for a facelift for the Ring, which thereafter looked an obviously sixties design. Two years ahead of the expiry of its 99 year lease on the property, the Exchange took the opportunity to move to Plantation House in Fenchurch Street. There it was possible to set up a modern purpose-built trading floor with adjacent Board Room and Visitors' gallery. Of functional importance was the ring of phone booths around the Ring, considerably improving communication between the Ring and member firms' offices.

At the time there were strong blandishments for the LME to join with other London futures markets, notably softs and the IPE, in a single futures centre with some common services. At the time the softs markets were not in the best of health and the property aspect of the proposal was not cut and dried either, so the LME's decision for an independent home was a logical one. This piece of recent history is worth remembering, however, as proposals are again circulating as this book goes to press for a convergence of the LCE and IPE on new combined premises, perhaps even also involving Liffe. With a hidden agenda behind this talk of something more than co-habitation, and powerful US markets casting envious eyes on the faster-growing futures markets in Europe, the LME may well find itself in future once again in a climate of instability for futures markets, notwithstanding its own very sound position both in its present home, and in operational terms.

It was not expected that the Plantation House address would provide the astonishing stability of Victorian Whittington Avenue and nor did it do so. In 1994 the market decamped once more to a new purpose-built trading room at 56 Leadenhall Street where sufficient space is available for a considerable expansion of the administrative function essential for an active and trade-oriented market. Interestingly, the total number of seats that can be provided round the Ring is still less than the 40 that constituted the framework of stability for Whittington Avenue.

PEOPLING THE MARKET

Like most service activities, futures markets are shaped by the people that constitute the market, both professionals and clients. As far as LME personnel are concerned, there have been several notable phases in its evolution. We have already remarked that before World War I there was a strong German presence on the Exchange, in line with Germany's central role in international metal commerce. Not surprisingly, after the war there was reaction against this, and the Exchange became notably "British" in its composition and outlook, whilst still serving a global clientele. The rules about British domicile of members were strictly enforced, including requirements for a majority of board members of such firms to be British nationals.

Not surprisingly, this attitude was again strongly in evidence after World War II, but by the sixties was beginning to break down under continuing globalisation of the Exchange's business. Also influential were individual cases of influential and congenial foreigners who had been so long resident in the UK they were hardly detectable as such. Exceptions began to be made.

Shortly after this a further leavening influence came from a less-expected quarter. Successful firms on the softs futures markets cast envious eyes on the buoyant and rather less trammelled business being done on the LME, and there was a small wave of new entrants to the Ring, and acquisitions of established LME firms, from this direction. This was known as the "Mincing Lane Raid". One of its far-reaching effects was the rise of one trader from this background to be chairman of the Committee. Whilst scandalising his conservative contemporaries by canvassing the merits of changing the LME to a cleared market, he also managed in a short few years to introduce the nickel and aluminium contracts to the LME's range of activities.

Further liberalisation of both personnel and parent companies' domicile continued in the seventies with increasing Continental and US ownerships and presences. However, operating companies continued, of necessity, to be UK domiciled and managed. The next wave of newcomers was from Japan, with the four biggest trading houses in non-ferrous metals gaining an interest in Ring dealing firms within a few years of each other, though on rather different formulae, not least in response to the LME's own changes in the criteria it imposed during this phase.

All this cumulative change was increasingly peopling the Ring with representatives of trade interests – major merchant houses and producers – though this was somewhat balanced by the presence of US commission houses. But it is a fact that, at the time of the tin crisis and the contemporary drastic re-structuring of the Exchange, only one firm was left on the LME which typified a prewar broker – namely a privately-owned partnership. This same change has seen the character of the faces around the Ring also change completely.

When trading resumed after World War II, the typical Ring dealer was a director or partner of his firm, well able to take major trading decisions on the spot. As the firms themselves began to change and new generations of management took over, people at that level increasingly stayed close to the back office and the client telephones. They entrusted Ring dealing to young professionals – often unkindly referred to as "the barrow boys" – with sharp reflexes, leather lungs and a willingness to operate within narrow limits laid down from the office. Today the LME Ring in action has more similarities than ever before with rings or pits on other futures exchanges around the world. Last but not least, the sexes are now equally represented, yet it is less than twenty years since the appointment of the first female authorised clerk caused a sensation.

APPENDIX I

Glossary of Commodity Market Terms

American option: An option, normally not used for futures contracts, which can be declared and the underlying metal transferred as a spot transaction for cash, at any time prior to expiration.

Arbitrage: An operation which involves a purchase of metal in one market with the simultaneous sale of an equivalent quantity of the same metal in another market, (eg the LME and The New York Commodity Exchange), and any necessary foreign exchange transaction to protect against any change in the parities if two currencies are involved.

Asian option: A cash-settled option on the average price of a given metal in a given averaging period.

Assay: The independent evaluation of the physiochemical composition of a metal to determine its degree of purity. LME-registered brands must meet specified metal contents and maximum levels of impurity, and are assayed to establish this.

At-the-money: An option whose strike price (qv) is the same as the current market price of the underlying metal.

Backpricing: A now almost obsolete facility whereby a consumer can have the option of 'fixing' a proportion of his period pricing intake on the previous day's settlement.

Backwardation: The situation when the cash or spot price of a metal is greater than its forward price. A backwardation occurs when a tight nearby situation exists in a metal. The size of the backwardation is determined by differences between supply/demand factors in the nearby positions compared with the same factors on the forward position. There is no inherent limit to the backwardation. The backwardation is also referred to as the 'back'.

Base Metal: Major industrial non-ferrous metals other than precious metals and minor metals.

Basis Price: The price agreed between the seller and the buyer of an option (qv) at which the option can be exercised. This price is also called the 'strike price'.

Bear: One who believes prices will move lower.

Bear Covering: The closing of a short position (qv).

Bear Market: A market in which prices are declining.

Bear Position: One where the trader has sold metal for forward delivery with the view to buying back at a lower price, his profit being the difference between the sale and repurchase price.

Bear Spread: An option spread (qv) of either puts (qv) or calls(qv) whereby the holder of the position benefits when prices fall.

Best Orders: Orders to buy or sell which the LME Ring-dealing broker executes on the market at what the dealer judges to be the best available price. It is also termed buying (or selling) 'at best'.

Bid: The price the buyer is prepared to pay. (Also see Offer).

Black-Scholes: A formula (named after its two inventors) which relates the premium value of an option to a combination of the value of the underlying metal, the time to expiry, the volatility of the metal and the prevailing interest rate.

Borrowing: Derived from 'borrowing metal from the market', which is achieved by buying a nearby date and simultaneously selling a date further forward. (See also Cash and Carry).

Broker: One who puts buyer and seller in touch with one another for a fee. In a broader sense the term is used for an organisation which is a member of an exchange and which acts as the means through which clients have access to the market. In these circumstances the broker acts as principal to his client but does not share in the risk as each deal is implemented by trading with other brokers in the market.

Brokerage: Commission charged by a broker for completing a transaction on the exchange.

Bull: One who believes prices will rise.

Bull Market: A market in which prices are rising.

Bull Position: One where the trader has bought metal with the view to selling it later after prices have risen to higher levels, his profit being the difference between the purchase and sale prices.

Bull Spread: An option spread (qv) of either puts (qv) or calls (qv) whereby the holder of the position benefits when prices rise.

Business Day: Any day except Saturday, Sunday or public/bank holiday in England or a day designated not to be by the LME directors.

Butterfly spread: An option strategy combining two option spreads (qv) whereby two options are bought (or sold) at the same strike price (qv) and one option is sold (or bought) at each of an equidistant higher and lower strike price, all for the same expiration.

Buyers over: If a bidder remains unsatisfied at the end of any ring, there are said to be buyers over. If this is so at the end of the second morning ring, the bid price will be the official buyers' price for that position.

Calendar Spread: An option spread (qv) whereby a pair of calls or puts for the same strike price (qv) but different expiration (qv) is traded simultaneously, one bought and one sold.

Call Option: An option which gives the buyer the right, but not the obligation, to purchase a particular futures contract at a specific strike price (qv) for a specific expiration (qv).

Carrying: Carrying is the general term used for both borrowing (qv) and lending (qv) on the LME. It is also used to describe a borrowing operation which results in physical metal being held in an LME warehouse by the borrower.

Cash: On the LME, the spot position for trading (with settlement on the second following day). On other futures markets, cash usually refers to physical (qv).

Cash and Carry: When a contango (qv) exists, the premium of the forward position over the prompt generally reflects costs of storage, insurance and finance for that period. When metal is in surplus the contango may widen to the point where an effective interest at more than money market rates can be locked in. (See also Borrowing).

Cathode: A flat rectangular piece of metal which has been refined by electrolysis. Copper and nickel are commonly traded and delivered in this form. Copper is always traded as whole plates. Nickel may be whole plates or cut into squares of various sizes down to 1⊕x1⊕.

Clearing: The process of settling all the differences between Ring members at the end of the day's trading. It is implemented through the clearing house (qv).

Clearing House: An independent body appointed or owned by an exchange to clear and guarantee business transacted between member brokers.

Clerk: An assistant to an LME Ring dealer. He or she may be 'authorised' (empowered to deal in the absence of the dealer) or 'unauthorised' (empowered only to record and check transactions).

Comex: The New York Commodity Exchange, on which copper, gold and silver are traded. Comex is often referred to as meaning the prices of these metals in New York, eg 'Comex is 90c', meaning the Comex copper price is 90 cents a lb Nymex – the New York Mercantile Exchange – with which Comex is merged trades platinum futures.

Commission: The fee paid by the client to the broker for executing the transaction.

Commission House: A term referring to brokerage houses, usually American, which execute general commissionable business on behalf of clients in securities and futures.

Contango: The situation when the price of a metal for forward or future delivery is greater than the cash or spot price of the metal. Contangos occur when the metal is in plentiful supply. The size of the contango does not normally exceed the cost of financing, insuring and storing the metal over the future delivery period. (See also Cash and Carry).

Corner: A position where one operator owns all or virtually all the market stocks of a commodity. His object is to control the supply of the metal, and thus cause an exceptional rise in price. The LME has rules to avert the worst effects of a corner. (See also Squeeze).

Cover: The balancing of an open position by buying (if the original position was short) or selling (if the original position was long) on the market.

CTA: Commodity Trading Advisor – a firm or more likely an individual who uses charts and system trading to trade on futures markets on behalf of managed investment funds.

Custom Smelter: A smelter which depends for its intake mostly on concentrate purchased from independent mines and on scrap metal, rather than its own captive mine sources.

Day Order (GFD): An order placed with a broker which is valid throughout the rings and kerbs during the day on which it was placed. If not executed that day, it is automatically cancelled. If valid for all markets (ie premarket, lunchtime and evenings as well as the rings and kerbs), any instruction should clearly state 'all markets'. The order can then be executed at any time of the day.

Declaration: The act of an option buyer informing an option grantor that he is taking up his right to buy (or sell) metal under the option contract.

Declaration Date: The date on which the buyer's right to exercise his option (qv) expires if not declared.

Delivery Date or Prompt Date: the date on which a metal has to be delivered to fulfil the contract terms.

Deposit: See Initial Margin.

Difference: In clearing (qv), contracts are concluded at maturity by the payment by buyer to seller (or vice versa) of the difference between the contract price and the corresponding settlement price.

Double Option: An option – sometimes known as a straddle (qv) – which gives the buyer or taker of the option the right either to buy from or sell to the seller (or grantor) of the option at the basis price.

EFF: Exchange for futures – a transaction whereby an off-Exchange futures contract is exchanged for an on-Exchange one.

EFP: Exchange for physical – a term originating on American futures exchanges but now increasingly in common use on the LME whereby two physical counterparties swap a physical parcel and an LME position at mutually agreed prices to avoid each of them having to hedge the transaction separately.

Electrolysis: A process by which direct current passes from one electrode (the anode) through a liquid electrolyte to another electrode (the cathode) in a cell. The passage of ions from the anode causes the pure metal only to be electrodeposited on the cathode. Because of the purity of the deposited metal, electrolysis is often used in refining. Impurities fall to the bottom of the cell as a slime and may contain valuable by-products.

Electrowinning: An electrolytic refining process in which the metal is recovered from the electrolyte rather than from an anode.

End Consumer: A company in whose hands the identity of the fabricated metal is finally lost in a more complex product, eg a refrigerator, motor car or transformer. Some of these companies also have fabricating subsidiaries. (See also Fabricator).

European option: An option which in practice is only declared on the expiration date because the underlying metal can only be acquired at that time. The LME option contract, which in theory can be declared at any time prior to expiration, is effectively a European option because there is no economic benefit to early declaration.

Expiration Date: The date at which the buyer's rights under an option contract expire unless declared (see Declaration Date).

Fabricator: A company which transforms refined metal (and sometimes scrap as well) into semi-fabricated products, (eg wire, cable, tubes, strip, rods) for sale to an end-consumer. Also referred to as a semi-fabricator.

Force Majeure: The clause in a metal supply contract which allows the seller not to deliver or the buyer not to take delivery of the metal, concentrate or scrap under the contract because of events beyond his control. There is no force majeure clause in an LME contract. Customers affected by a declaration of force majeure by a producer or refiner can always turn to the LME as a source of metal, and suppliers can deliver their metal to the LME if their customers declare force majeure.

Forward: The purchase or sale of metal for delivery at a specified future date. Hence Forward Price, Forward Contract.

Give-ups: A give-up occurs when a client instructs an executing LME broker to give up an uncleared contract to another of his LME brokers for the purpose of being entered into the clearing and eventually settled. (See also Transfer).

Good Till Cancelled Order (GTC): An order to buy or sell which is valid every market day during rings and kerbs (or during the entire day if specified), until cancelled by execution or client's instructions.

Grantor (of an option): The person or firm that sells an option.

Hedging: The establishment of an opposite position on a futures market or by means of options from that held and priced in the physical commodity. Without hedging, the physical position would be at risk to price fluctuations.

Initial Margin: The initial outlay of money required to open a futures position. The size of this deposit varies from metal to metal, but is usually a fixed amount, set by the Clearing House (qv), representing a small percentage (around 5%) of the value of the contract. (See also Variation Margin).

In Warehouse: Title to metal sold under LME contracts passes with delivery of the warehouse warrant (qv). LME prices therefore refer to metal lying in LME warehouse.

Integrated Producer: A producer of metal who owns mines, smelters and refineries, and sometimes also fabricating plants.

In-the-money: An option whose strike price (qv) is above the current market price if a put or below it if a call.

Introducing broker: Sometimes called a half-commission man; an individual who introduces, and usually services, a client to a clearing member of the LME.

Inter-Market: Dealings in a metal between brokers either "round the back" of the Ring, over the telephone or on screen, which take place between rings and kerbs or in the period between the close of the morning kerb and the first afternoon Ring.

Inter-Office: Deals concluded between members as official LME business at any time outside Ring trading hours.

Kerb Trading: This is a period of fifteen to twenty five minutes at the end of each session (am and pm) where all metals are traded simultaneously around the ring. Kerb trading is often expressed simply as'on the kerb' or 'kerb'. The term comes from the early history of commodity futures markets when trading after hours was literally conducted on the kerb of the street.

Late Night Trading ('Lates'): The period of unofficial LME trading from the close of the afternoon kerb to the close of the comparable US markets.

Lending: Derived from lending metal to the market. Selling metal on a nearby date and simultaneously buying it back on a forward date, thereby extending a long position.

Limit: In some futures markets (but not on the LME) there is a limit to the rise or fall in the price of a commodity during a given period of time. Once this limit is reached, trading is suspended until the time period expires, when trading can resume. The suspension relates only to forward contracts, the spot month continuing to trade.

Limit Order: An order placed by a client to buy or sell at a specified price.

Liquidation: Closing out of a long position. It is also sometimes used to denote closing out of a short position, but this is more correctly referred to as covering or covering in.

Long: An open purchased futures position. Thus 'to go long' means starting a transaction by the purchase of a futures contract. Equally a producer or a processor may be long of physical metal if his supply of metal exceeds his sales orders.

Lot: The minimum amount of a commodity in which one may deal on a futures market. On the LME a lot is often called a warrant or contract.

Margin Call: See Variation Margin.

Mark to Market: Evaluating an open position on the basis of current market price, usually to assess the need for a variation margin (qv).

Market: a) The location of trading.

b) The current price of a commodity.

Market Order: An order to buy or sell a futures contract at the first obtainable price or prices on the market for tonnage involved. If the order is given "with discretion" the broker may execute it over time as he sees fit; if "without discretion" he must hit every bid or offer as it appears until the order is filled.

Market if Touched: or MIT: A term usually used by chartists. It is an instruction to buy or sell at market if a certain level above or below the current price is reached.

Matching System: The system operated by the Clearing House (qv) for the matching and confirmation of Exchange contracts between clearing members.

Merchant: A metal merchant, as distinct from a producer's agent or broker, often acts as a principal, buying metal or concentrate from producers and others and selling it on to others. He will often hold metal on his own account while waiting for a buyer.

Min-max: An option strategy involving selling an option (eg a put option) to fund the purchase of an opposite option (ie a call option).

Nearby: A date/contract that is relatively close to the cash/spot/expiry date.

Nominal Price: An estimate of the price for a future month date which is used to designate a closing price when no trading has taken place in that date. Also used for current price indications in similar circumstances in physical trading.

Novation: The technical term for the process whereby trades between Ring dealers are guaranteed by the clearing house (qv).

Nymex: New York Mercantile Exchange. (See also Comex).

Offer: The price the seller asks for the commodity on offer. (See also Bid).

One Cancels Other (OCO): A trading instruction incorporating two orders with a built-in further instruction to cancel one if the other is executed. The order can indicate two different price levels or a combination of price and time limits.

Open Outcry: the method of dealing employed on the LME and other futures markets. A broker announces to the entire ring the amount of a commodity he has for sale (or seeks to buy), and the required delivery date and price. When another broker responds to accept the bid or offer, the deal is fixed and that price becomes the latest traded price for that delivery date.

Open Position: A forward market position which has not been closed out. For some exchanges the net open position is an important statistic and is published daily.

Option: An option is a contract which gives the buyer the right but not the obligation for a specified period of time to buy from (call) or sell to (put) the grantor a specified quantity of metal at a specified strike price on a specified delivery date in return for payment of a negotiated premium. The option must be declared by the buyer on or before its expiration date.

Option Spread: The simultaneous purchase and sale of a pair of calls (or puts) in the same metal but at different strike prices or different expiration dates, or both. The option equivalent of a carry (qv).

Out-of-the-money: An option whose strike price (qv) is above the current market price if a call and below it if a put.

Path-dependent Options: Options which are automatically granted, exercised or abandoned if certain specified conditions occur in the future. Examples are Knock-in, Knock-out and Look-back options; these are not much used in the metal industry.

Period pricing: A contract for physical metal supply over a period (usually a year) in which all elements except price are settled in advance. Invoice prices are then structured on, eg, the average LME price for that metal for the month of delivery.

Physical: Trade in the commodity itself, ie not hedging (qv), speculation or arbitrage (qv).

Premarket: Trading over the telephone among brokers before the LME opens for ring-dealing. Premarket trading can be very active, influenced by the hedging operations of producers and consumers.

Premium: (i) The cost of buying an option. (ii) The amount by which the price for one delivery date is greater than that for another date. (iii) The difference in price between different refined products, eg between ingot and billet.

Price (used as a transitive verb): To fix the price of a purchase or sale on one or more LME quotations, usually the official settlement price for the metal concerned.

Pricing-in (out): When a merchant agrees to buy a quantity of metal from a consumer both parties will agree in advance that the price basis will be the closing price of a specific ring or rings on the LME. The buyer will therefore be pricing-in on the close of that ring, and will probably hedge his purchase by selling during the closing stages of that ring. The consumer will be pricing-out and may hedge by buying at the close.

Principal: A person or organisation able to make, and obliged to fulfil, all the conditions of a contract.

Prompt: For immediate (ie two days) delivery.

Prompt Date: See delivery date.

Put Option: An option which gives the buyer the right, but not the obligation, to sell a particular futures contract at a specific strike price (qv) for a specific expiration (qv).

Refinery: A processing plant which produces high purity metal either by electrolysis, electrowinning or fire-refining. In copper and lead production, refining is always preceded by smelting (qv), but for tin, zinc and nickel, smelting and refining blend into one process of recovering marketable quality metal from concentrates. Refiners (and smelters) may also treat scrap materials. In aluminium, a refinery produces alumina.

Ring: a) A five-minute trading period in a single metal on the LME.

b) The circle of seats where such trading takes place.

Ring Dealing Broker: An organisation which is a full clearing member of the LME and entitled to deal in the Ring. Some clearing members do not have Ring dealing privileges.

Risk Management: The management of metal price risk on an ongoing basis by the use of LME futures and options with a view to containing the impact of adverse price movements and enhancing profitability. The concept is an extension of the more simple type of hedging whereby the pricing of specific transactions is offset or fixed.

Roll: To transfer a market position in futures or options from one date (forward or backward) or one strike price (up or down) to another by a simultaneous purchase and sale.

Secondary Metal Dealer: A firm which specialises in buying and selling scrap metal in bulk. The metal is often in alloy form (such as brass) and often in a fabricated form, but is always at risk from fluctuations in the value of the basic metal.

Sellers over: The opposite of buyers over (qv).

Semi-fabricator: See Fabricator.

Semis: Usual trade term for semi-fabricated products (sheet, wire, tube etc).

Session: There are two sessions on the LME each day, one in the morning and one in the afternoon. Each session comprises two five-minute 'rings' for each metal and a fifteen to twenty-five minute 'kerb' period.

Settlement Price: The last unfulfilled offer to sell cash at the close of the second morning ring in each metal – ie the cash seller's price at that close. This price, which is officially announced on the LME floor, prevails as the accepted cash reference price for the metal for the succeeding twenty four hours.

Short: An open sold position on a futures market. Thus 'to go short' is to start a transaction by the sale of a futures contract. Equally a producer or processor may be short of physical if his supply of metal is less than his sales orders.

Smelter: A processing plant which produces crude metal by treating mine feed (concentrate) or residues. In the case of aluminium a smelter produces refined metal from alumina.

Spot Month: The first deliverable month for which a quotation is available on a futures market. For example, 'March is now the spot month on Comex'. As the LME trades in daily dates, it does not have a spot month.

Squeeze: Pressure on a particular delivery date which makes the price of that date higher in relation to other dates.

Stockist: A stockist holds supplies of semi-fabricated products ready for sale to users. Stockists usually specialise either in steel/stainless steel/aluminium or in copper/brass/high-value metals. A stockist function is also fulfilled by, eg a builders' merchant holding large stocks of copper tubing or plumbing fitments. (See Fabricator).

Stop-Loss Order: An order which becomes a market order to buy only if the market advances to a specified level or to sell only if the market declines to a specified level. As soon as this specified level is traded the order is executed for the client at the next obtainable price. There is no guarantee the order will be executed at the price specified. A stop-loss order, as its name implies, is instituted to prevent or minimise losses from either a short or long position.

Straddle: The simultaneous purchase (or sometimes sale) of put and call options at the same strike price. Normally used when an increase in volatility is expected.

Strangle: The simultaneous sale (or purchase) of out of the money calls and out of the money puts for the same date.

Strike Price: See Basis Price.

Synthetic Put (or Call): A combination of a short (or long) futures position and a long out of the money call (or put) which protects the hedger against prices moving in the direction he fears, but also allows him to participate in part of a favourable move.

Swap: An agreement between a hedger and a broker or bank to swap a series of forward cash flows based on selling and buying at an agreed pair of prices. These may be fixed (a known LME price) or floating (a yet-to-be-determined LME average price). The instrument originated in the banking business but is similar to an LME averaging agreement whereby a broker agrees to guarantee a hedger an average price for opening or liquidating a hedge.

Switching: a) On the LME, the exchange of metal in one warehouse for that in another, eg Rotterdam for London.

b) The movement out of one futures contract into another, usually further forward.

Tender: Delivery of the physical commodity against a futures contract.

Tom-Next: Within the LME's normal two day settlement, a position may nevertheless have a prompt date one day forward. It can still be officially traded up to the first ring of that day under a facility known as Tom(orrow)-Next(day). This permits such positions to be rolled forward to the official cash date on payment of a negotiated premium.

Traded Option: An option contract based on standardised terms which can be traded to third parties or on an Exchange. The standard LME options contract is a traded option contract.

Transfer: Sometimes erroneously called give-ups, transfers are a process whereby a client instructs one of his LME brokers to transfer responsibility for an open cleared position to another of his brokers for the sake of strengthening his open position with the latter broker.

Turnover: See Volume.

Value: An LME term referring to a price which is traded in a volume sufficient to satisfy all the current buyers and sellers at that price.

Variation Margin: A request for additional funds (a margin call) to cover any losses on an open position, based on current market values. Variation margin is assessed and called for on a daily basis. An open position which is in profit against the current market has a positive margin, which can be used in assessing a client's credit position. On some US markets positive margin is paid out to the client.

Volume: The total weight of a commodity traded in one day, expressed either in tonnes or lots. Also referred to as turnover.

Warrant: The document of title to metal stored in an LME registered warehouse. The warrant is a bearer instrument and states the brand of metal, its weight, the number of pieces and the rent payable.

APPENDIX II

Special contract rules for metals

SPECIAL CONTRACT RULES FOR HIGH GRADE PRIMARY ALUMINIUM

1. QUALITY

Aluminium deliverable under this contract shall be:

(a) primary aluminium of minimum 99.70% purity with maximum permissible iron content 0.20% and maximum permissible silicon content 0.10%, or

(b) primary aluminium with impurities no greater than in the registered designation P1020A in the Registration Record of Aluminium Association Designations and Chemical Composition Limits for Unalloyed Aluminium of the Aluminium Association Inc., U.S.A. (May 15th, 1982);

(c) of brands listed in the LME-approved list of aluminium brands.

2. SHAPES AND WEIGHTS

Aluminium delivered may be in the form of ingots, T-bars or sows and ingots shall be securely strapped in bundles suitable for stacking. Ingot weight shall be within the permitted range 12kg to 26kg each. T-bar weight shall not exceed 5% more than 675kg each, and the weight of each sow shall not exceed 5% more than 750kg.

3. SIZE OF LOT

25 tonnes

4. WARRANTS

4.1 Warrants shall be for 25 tonnes each (2% either more or less).

4.2 Each parcel particularised in each Warrant shall comprise either ingots, T-bars or sows, shall lie at one warehouse, be the production of one country and shall consist of aluminium of one brand shape and size subject, in the case of ingots, to the necessity of including different shapes and sizes at the bottom of each parcel for the purpose of palletisation. Each parcel of ingots placed on warrant shall be delivered securely strapped in bundles not exceeding two tonnes each.

4.3 Warrants must contain the warning regarding entrapped moisture referred to in the Special Rules for Placing Aluminium on Warrant.

5. MAJOR CURRENCY

US dollars

SPECIAL RULES GOVERNING THE PLACING OF HIGH GRADE PRIMARY ALUMINIUM ON WARRANT

1. Each delivery of aluminium for placing on warrant shall be accompanied by a Certificate of Origin and a producer's Certificate of Analysis both of which must be lodged with the warehouseman. The Certificate of Analysis shall be within the quality specification set out in Special Rule 1 for the relevant grade of aluminium as above set out. A Bulk Analysis Certificate or a copy thereof is acceptable but if a producer's Certificate is not available an Analysis certificate must be prepared and signed by an assayer on the list of approved assayers. All analysis certificates must show each heat number.

2. The party intending to place on warrant aluminium for delivery in satisfaction of any Contract must advise the warehouseman of the name of a Ring Dealing or a Clearing Associate member of the London Metal Exchange who will accept responsibility for the initial placing of the Warrants on the market. Such Warrant(s) must be issued to the order of the said member who, prior to its or their endorsement, must be satisfied that the documentation is evidence of good delivery.

3. Each Warrant shall be made up from the production of one country and shall consist of one brand which is listed as being a good delivery. Each Warrant shall state the total weight of the parcel particularised thereon, its country of origin, brand and the date(s) and reference number(s) of the Certificates of Origin and of Analysis lodged with the warehouseman. Each Warrant shall be of 25 tonnes (2% either more or less) and shall state the number of bundles or T-bars or sows and in the case of bundles the number of ingots making up each bundle. In the case of either T-bars or sows the Warrant shall state the piece weight of each.

4. The warning clause set out below must appear on the Warrant:

> WARNING The buyer is advised that this metal may contain crevices and hidden recesses holding entrapped moisture. The metal should be handled and processed with this possibility in mind. Entrapped moisture may cause an explosion if the metal is introduced into a melting-furnace without proper drying.

5. On request, the warehouseman is obliged to submit to the holder of the Warrant Certificates of Origin/Analysis or copies thereof if the 25 tonnes parcel forms a part of a larger delivery covered by Bulk Analysis.

SPECIAL CONTRACT RULES FOR ALUMINIUM ALLOY

1. QUALITY

The aluminium alloy delivered under this contract shall be:

(a) aluminium alloy conforming to one of the specifications listed below:

(1) A380.1 produced in conformity with The Aluminum Association Inc. specification (1989);

(2) 226 produced in conformity with GBD-AlSi9Cu3 as described in DIN standard 1725 (1986);

(3) D12S produced in conformity with JIS H2118-1976, Class 12; (Note: this specification to be read in conjunction with the provision that there be an allowance as follows: Others, total 0.50% max., Al remainder);

(b) of brands listed in the LME-approved list of aluminium alloy brands.

2. SHAPES AND WEIGHTS

Aluminium Alloy delivered must be in ingot form and shall be securely strapped with galvanised or protectively-coated steel or polypropylene band in 500 to 1,000 kg bundles of equal weight suitable for stacking. Ingot weight shall be within the permitted range 4 to 14 kg each, with all ingots in any one lot being of the same weight and size. Ingots shall not exceed 800mm length - the only exception to be the cast feet on which bundles may be stacked.

3. SIZE OF LOT

20 tonnes

4. WARRANTS

4.1 Warrants shall be for 20 tonnes each (2% either more or less).

4.2 Each parcel particularised in each Warrant shall comprise ingots, shall lie at one warehouse, be the production of one plant and shall consist of aluminium alloy of a single aluminium alloy type. Each parcel of ingots placed on warrant shall be delivered securely strapped in bundles not exceeding 1,000 kg each. Each parcel shall contain aluminium alloy derived from no more than five melts, and each bundle shall contain aluminium alloy from one melt.

4.3 Warrants must contain the warning regarding entrapped moisture referred to in the Special Rules Governing the Placing of Aluminium Alloy on Warrant.

5. MAJOR CURRENCY

US dollars

SPECIAL RULES GOVERNING THE PLACING OF ALUMINIUM ALLOY ON WARRANT

1. Each delivery of aluminium alloy for placing on warrant shall be accompanied by a Certificate of Origin. A producer's Certificate of Analysis shall be provided for each melt represented within the lot and shall contain the melt number and melt date. Each bundle shall be tagged with a die-stamped or indelible tag stating the aluminium alloy type, melt numbers, melt date and the brand name. The Certificates of Analysis shall be within the quality specification set out in Special Contract Rule 1 for the relevant aluminium alloy type.

2. Aluminium Alloy delivered for warranting must be in lots of 20 tonnes, comprising bundles of the same size between 500 and 1,000 kg each. Aluminium Alloy can only be placed on LME warrant within eight weeks of the melt date. In the case of lots comprising several melts the earliest melt date shall determine eligibility for warranting.

3. Aluminium Alloy must arrive at warehouse in covered transport and must be unloaded in covered or dry conditions.

4. The party intending to place on Warrant aluminium alloy for delivery in satisfaction of any Contract must advise the warehouseman of the name of a Ring Dealing or an Associate Clearing Member of the London Metal Exchange who will accept responsibility for the initial placing of the Warrants on the market. Such Warrant(s) must be issued to the order of the said member who, prior to its or their endorsement, must be satisfied that the documentation is evidence of good delivery.

5. Each Warrant shall be made up from the production of one plant and shall consist of one brand which is listed as being good delivery. Each Warrant shall state the total weight of the parcel particularised thereon, its country of origin, brand and the date(s) and reference number(s) of the Certificates of Origin and of Analysis lodged with the warehouseman. Each Warrant shall be of 20 tonnes (2% either more or less) and shall state the number of ingots making up each bundle.

6. The warning clause set out below must appear on the Warrant:

> WARNING The buyer is advised that this metal may contain crevices and hidden recesses holding entrapped moisture. The metal should be handled and processed with this possibility in mind. Entrapped moisture may cause an explosion if the metal is introduced into a melting-furnace without proper drying.

SPECIAL CONTRACT RULES FOR COPPER - GRADE A

1. QUALITY

1.1 The copper delivered under this contract must be electrolytic copper in the form of either

(a) cathodes – Grade A, or

(b) wirebars – Grade A of standard dimensions in the weight range 110 kgs to 125 kgs

in sellers's option.

No further deliveries of the currently Listed Brands of Wirebars will be accepted for warranting after 31st December 1993.

Once the stocks of Wirebars as at 31st December 1993 are exhausted, the option to deliver Wirebars against the Grade A Contract will be withdrawn.

1.2 All copper delivered must be of brands listed in the LME-approved list of Copper – Grade A brands and must conform with BS 6017-1981 (1989) (cathode grade designation Cu-CATH-1, wirebar designation Cu-ETP-2).

2. SHAPES AND WEIGHTS

Each parcel of 25 tonnes shall lie at one warehouse and be of one brand, shape and size, but, in the case of cathodes, shall be of full or cut plates subject to the necessity of including different shapes and sizes at the bottom of each parcel for the purpose of palletisation. Cut plates shall not be less than quarter size. Each parcel of copper cathodes placed on warrant shall be delivered securely strapped in bundles not exceeding 4 tonnes and with the brand shown on the strapping. In the case of wirebars each parcel shall consist of wirebars of uniform piece weight subject to the usual tolerances. On and after 1st June 1985 each parcel of wirebars placed on warrant shall be delivered securely strapped, in bundles not exceeding 2.5 tonnes. Warrants issued prior to 1st June 1985 for unstrapped wirebars shall constitute a good delivery.

3. SIZE OF LOT

25 tonnes

4. WARRANTS

4.1 Warrants shall be for 25 tonnes each (2% either more or less).

4.2 Each cathode warrant must state the total weight of the parcel, the brand, the name of the producer, and the number of plates making up each parcel. Warrants for cathodes issued on and after 1st September 1981 must also state the number of bundles making up each parcel. Warrants issued after 5th November 1992 need not show the name of the producer. Warrants issued after 8th April 1993 need not show the number of plates.

4.3 Each wirebar warrant must state the total weight of the parcel, the brand, the wirebar piece weight and the number of wirebars making up each parcel. Warrants for wirebars issued on or after 1st June 1985 must also state the number of bundles making up each parcel.

4.4 Each warrant must be identified as Copper – Grade A.

5. MAJOR CURRENCY

US dollars

SPECIAL CONTRACT RULES FOR STANDARD LEAD

1. QUALITY

The lead delivered under this contract must be refined pig lead (minimum 99.97% purity). All lead delivered must be:

(a) of brands listed in the LME-approved list of lead brands;

(b) in pigs weighing not more than 55 kgs each.

2. SHAPES AND WEIGHTS

Each parcel of 25 tonnes shall lie at one warehouse and be of one brand and shall consist of pigs of one size, subject to the necessity of including different shapes and sizes at the bottom of each parcel for the purpose of palletisation. On and after 6th May 1992 each parcel placed on warrant shall be delivered securely strapped, in bundles not exceeding 1.5 tonnes. Each parcel placed on warrant between 1st June 1985 and 5th May 1992 shall be delivered securely strapped, in bundles not exceeding 1.2 tonnes. Warrants issued prior to 1st June 1985 for unstrapped pigs shall constitute a good delivery. Additionally, on and after 16th October 1995, each bundle placed on warrant shall be securely strapped in bundles to permit safe handling without bundle distortion and breakage.

3. SIZE OF LOT

25 tonnes

4. WARRANTS

Warrants shall be for 25 tonnes each (2% either more or less). Each Warrant must state the total weight of the parcel, the brand, and the number of pigs making up each parcel. Warrants issued on or after 1st June 1985 must also state the number of bundles making up each parcel.

5. MAJOR CURRENCY

US dollars

SPECIAL CONTRACT RULES FOR PRIMARY NICKEL

1. QUALITY

The nickel delivered under this contract must be Primary Nickel of minimum 99.80% purity with chemical analysis conforming to the current ASTM specification.

2. SHAPES AND WEIGHTS

2.1 All nickel delivered must be:

(a) of the production of those producers named in the LME-approved list;

(b) in the form of either cathodes or pellets or briquettes.

2.2.1 In the case of cathodes deliveries shall be made in the form of sizes cut to either 100 mm x 100 mm (4″x 4″), 50 mm x 50 mm (2″x 2″), or 25 mm x 25 mm (1″x 1″), size tolerances in accordance with internationally accepted trade practice.

2.2.2 Each warrant shall consist of only one size.

2.3 All nickel delivered shall be packed in sound steel drums with a nett weight of minimum 150 kgs and maximum 500 kgs. On or after 16th April 1995, all nickel delivered for warranting shall be packed in sound steel drums with an even nett weight (+/- 2% more or less) of minimum 150 kgs and maximum 500 kgs. Each warrant shall consist of drums of uniform size and weight. The gross and nett weights must be clearly marked/stamped on each individual drum together with the producer's or brand name.

2.4 Warehouse Inspection of Drummed Primary Nickel

2.4.1 Nickel in original producer drums with Ring Dealing or Associate Clearing Member guarantee:

2.4.1.1 Nickel in original producer drums with producer seal intact can be placed on LME Warrant without opening every drum under certain conditions, namely:

(a) that the Ring Dealing or Associate Clearing Member warranting the material certifies in writing to the warehouse that all the nickel in any one delivery is from one source known to that Member;

(b) that the certificate should identify the nickel with drum numbers or other markings specific to the certified volume;

(c) that the warehouse opens one drum in ten from each batch so certified. The one drum in ten is to be selected at random from the batch.

2.4.1.2 Provided this inspection confirms that the contents of the opened drums conform to the producer markings for both product and weight, the warehouse is permitted to enter the other drums into stock unopened.

2.4.1.3 The warehouse should mark each drum with its distinguishing mark or seal identifying the warehouse and indicating that the drums were entered into stock uninspected.

2.4.1.4 The warehouse will be required to retain the Member certification on record for a minimum of six months after withdrawal of the nickel from that LME listed warehouse. Any complaint which might arise after withdrawal can be taken up via the Member concerned, with the help of the warehouse, with the Member who provided the original certification.

2.4.2 All other nickel in drums:

All other drummed nickel to be placed on LME Warrant must, without exception, be opened and emptied on arrival by the LME listed warehouse and the nickel inspected, repacked and re-sealed, in original drums where possible. This includes:-

producer nickel in original producer drums where there is no written Member guarantee;
producer nickel in unmarked drums;
producer nickel in original drums which are in unsafe condition and or where the seals are broken;
repacked producer nickel;
producer nickel which has been withdrawn from LME listed warehouse and is subsequently presented for re-warranting.

2.4.3 The warehouse will be responsible for resealing all inspected drums using their own numbered seals. All costs incurred will be for account of the party instructing the warehouse to place the material on Warrant.

3. SIZE OF LOT

6 tonnes

4. WARRANTS

4.1 Warrants shall be for 6 tonnes each (2% either more or less). Each parcel of 6 tonnes shall be the product of one producer in one country, and shall consist of one shape and size and shall lie in one warehouse.

4.2 Each warrant must state the name of the producer, the shape, the dimensions of cathodes where applicable, the gross and nett weights and the numbers of the drum seals making up each parcel.

5. MAJOR CURRENCY

US dollars

SPECIAL NOTE ON PLACING NICKEL ON WARRANT

The party intending to place nickel on Warrant for delivery to the Market must advise the warehouseman of the name of a Ring Dealing Member or Associate Clearing Member who will be responsible for the initial placing of the Warrant(s) on the Market, and such Warrant(s) must be issued to the order of that Ring Dealing Member or Associate Clearing Member. The obligations of the Ring Dealing Member or Associate Clearing Member to whose order the Warrants are issued are in no way diminished by the implementation of the procedures in Regulation 2.4 above.

SPECIAL CONTRACT RULES FOR TIN

1. QUALITY

The tin delivered under this contract must be refined tin of minimum 99.85% purity and conforming to BS3252:1986. All tin delivered must be:

(a) of brands listed in the LME-approved list of tin brands;

(b) either in ingots or slabs each weighing not less than 12 kgs or more than about 50 kgs.

2. SHAPES AND WEIGHTS

Each parcel of 5 tonnes shall lie at one warehouse and be of one brand, shape and size, subject to the necessity of including different shapes and sizes at the bottom of each parcel for the purpose of palletisation. Each parcel placed on warrant shall be delivered securely strapped, in bundles not exceeding 1.2 tonnes.

3. SIZE OF LOT

5 tonnes

4. WARRANTS

Warrants shall be for 5 tonnes each (2% either more or less). Each Warrant must state the total weight of the parcel, the brand, and the number of ingots or slabs making up each parcel, and must also state the number of bundles making up each parcel.

5. MAJOR CURRENCY

US dollars

SPECIAL RULES GOVERNING THE PLACING OF TIN ON WARRANT

1. Each delivery of tin for placing on warrant shall be accompanied by a producer's Certificate of Analysis which must be lodged with the warehouseman. The Certificate of Analysis shall be within the quality specification set out in Special Contract Rule 1. A Bulk Analysis Certificate or a copy thereof is acceptable, but if a producer's Certificate is not available an analysis certificate must be prepared and signed by an assayer on the list of approved assayers.

2. Each Warrant shall be made up from the production of one country and shall consist of one brand which is listed as being a good delivery. Each Warrant shall state the total weight of the parcel particularised thereon, its country of origin, brand and the date(s) and reference number(s) of the Certificate of Analysis lodged with the warehouseman.

3. On request, the warehouseman is obliged to submit to the holder of the Warrant Certificates of Analysis or copies thereof if the 5-tonne parcel forms a part of a larger delivery covered by Bulk Analysis.

SPECIAL CONTRACT RULES FOR SPECIAL HIGH GRADE ZINC

1. QUALITY

The zinc delivered under this contract must be zinc of minimum 99.995%. purity. All zinc delivered must be:

(a) of brands in the LME-approved list of special high-grade zinc brands;

(b) either in slabs or in plates or in ingots weighing not more than 55 kgs each.

2. SHAPES AND WEIGHTS

Each parcel of 25 tonnes shall lie at one warehouse and be of one brand and shall consist of either slabs, plates or ingots of one size subject to the necessity of including different shapes and sizes at the bottom of each parcel for the purpose of palletisation. Each parcel placed on warrant shall be delivered securely strapped in bundles not exceeding 1.5 tonnes.

3. SIZE OF LOT

25 tonnes

4. WARRANTS

Warrants shall be for 25 tonnes each (2% either more or less). Each Warrant must state the total weight of the parcel, the brand, and the number of slabs, plates or ingots making up each parcel. Warrants issued on or after 1st June 1985 must also state the number of bundles making up each parcel.

5. MAJOR CURRENCY

US dollars

APPENDIX III

LME Brands

Primary Aluminium Brands

Official London Metal Exchange list of the brands deliverable in fulfilment of its HIGH GRADE PRIMARY ALUMINIUM contract:

		Producers
Argentina	ALUAR	Aluar Aluminio Argentino SAIC
Australia	ALCAN	Alcan Australia Ltd
	ALCOA OF AUSTRALIA	Alcoa of Australia Ltd
	BSL	Boyne Smelters Ltd
	COMALCO	Comalco Aluminium (Bell Bay) Ltd
	PORTLAND	Portland Smelter Services Pty Ltd
	TOMAGO	Tomago Aluminium Co. Pty Ltd
Bahrain	ALBA	Aluminium Bahrain BSC
Brazil	ALBRAS	Albras Aluminio Brasileiro SA
	ALCAN BRASIL	Alcan Aluminio do Brasil SA
	ALCOA BRASIL	Alcoa Aluminio SA
	CBA	Companhia Brasileira de Aluminio SA
	SAO LUIS	Alcoa Aluminio SA (ALUMAR)
	SAO LUIS	Billiton Metais SA (ALUMAR)
	VALESUL	Valesul Aluminio SA
Cameroon	ALUCAM	Alucam
Canada	ABI	Aluminerie de Becancour Inc.
	ALCAN	Alcan Aluminium Limited
	ALOUETTE	Aluminerie Alouette Inc.

	REYNOLDS	Canadian Reynolds Metals Co.
China	AL	Shandong Aluminium Corp.
	CNM&M	China National Metals & Minerals
	G*L	Guizhou Aluminium Smelter
Dubai	DUBAL	Dubai Aluminium Company Ltd
Egypt	EGYPTALUM	The Aluminium Company of Egypt
France	PECHINEY	Pechiney SA
Germany	HA	Hoogovens Aluminium HuttenwerkGmbH
	HAW	Hamburger Aluminium-Werk GmbH
	LMG	Leichmetall GmbH
	VAW	VAW Aluminium AG
Ghana	VA	Kaiser Aluminium & Chemical Corp.
	VALCO	Kaiser Aluminium & Chemical Corp.
Greece	ADG	Aluminium de Grece SA
Hungary	AJKA/HUNGALU*	Hungalu Aluminium Industrial Ltd
	INOTA/HUNGALU	Hungalu Aluminium Industrial Ltd
	TATABANYA/ HUNGALU**	Hungalu Aluminium Industrial Ltd
	TBALUMEO/ HUNGALU**	Hungalu Aluminium Industrial Ltd
Iceland	ISAL	Icelandic Aluminium Co. Ltd
India	BALCO	Bahrat Aluminium Co.
	INDAL	Indian Aluminium Co. Ltd
	HA	Hindalco Industries Ltd
	HINDALCO	Hindalco Industries Ltd
	NALCO	National Aluminium Co. Ltd
Indonesia	INAL	Pt. Indonesia Asahan Aluminium
Iran	IRALCO	Iranian Aluminium Co.
Mexico	ALUMSA***	Aluminio SA de CV
Netherlands	DELFZIJL	Aluminium Delfzijl BV
	PNL	Pechiney Nederland N V
New Zealand	NZAS	New Zealand Aluminium Smelters Ltd
Norway	HYDRO	Hydro Aluminium AS (Ardal)
	HYDRO	Hydro Aluminium AS (Karmoy)
	HYDRO	Hydro Aluminium AS (Hoyanger)
	HYDRO	Hydro Aluminium AS (Sunndalsora)
	SOERAL	Soer-Norge Aluminium AS
Poland	WR 36	Huta Aluminium "Konin" SA
Rumania	IAS	Alro SA Slatina
Russia	BAS	Bogoslovsky Aluminium Smelter (BAS)
	CAA3	Sayansky Aluminium Plant
	CCCP	c/o Razmpoimport (Production of Volgogradsky)

**not good delivery if placed on warrant after 17.10.92*
***not good delivery if placed on warrant after 17.6.93*
****not good delivery if placed on warrant after 1.1.94*

	HA3	Nadvoitsky Aluminium Smelter
	HKA3	Novokuznetski Aluminium Plant
	ID	Bratsk Aluminium Plant
	KPA3	Krasnoyarsk Aluminium Plant
	URKA3	c/o Raznoimport (Production of Irkutsky)
Slovakia	ZSNP	ZSNP AS Aluminium Works
Slovenia	TALUM	TALUM, d.o.o., Kidricevo
South Africa	ALUSAF	Alusaf (Pty) Ltd
Spain	ALUMINIO ESPANOL	Aluminio Espanol SA
	INESPAL	Industrial Espanola del Aluminio SA
Suriname	SURALCO*	Suriname Aluminium Co.
	SURALCOA	Suriname Aluminium Co.
Sweden	GRAENGES ALUMINIUM	Graenges Aluminium Metall AB
Tajikistan	TADA3	c/o Raznoimport (Production of Tajikisky)
Turkey	ETIAL 7	Etibank
UK	AAM	Anglesey Aluminium Metal Ltd
	ALCAN	Alcan Aluminium (UK) Ltd
	BACO	Alcan Aluminium (UK) Ltd
USA	ALCOA	Aluminium Co. of America
	ALUMAX**	Alumax Inc.
	CAC	Columbia Aluminium Corp.
	CFAC	Columbia Falls Aluminium Co.
	EASTALCO	Eastalco Aluminium Co.
	INTALCO	Intalco Aluminium Co.
	KAISER	Kaiser Aluminium & Chemical Corp.
	LAURALCO	Alumax Inc.
	MT HOLLY	**Alumax of South Carolina**
USA	NSA	Southwire Co.
	NORANDA ALUMINIUM INC	Noranda Aluminium Inc.
	ORMET	Ormet Corp.
	NWA	Northwest Aluminium Corp.
	REYNOLDS	Reynolds Metals Co.
	VANALCO	Vanalco Inc.
	-	Alcan Aluminium Co.
Venezuela	ALCASA	CVG Aluminio del Caroni SA
	VENALUM	Industria Venezolana de Aluminio SA
former	KAT	Kombinat Aluminijuma Titograd
Yugoslavia	MOSTAR***	Energoinvest Mostar

**not good delivery if placed on warrant after 11.8.94*
***not good delivery if placed on warrant after 8.3. 94*
**** not good delivery if placed on warrant after 3.5 94*

Aluminium Alloy Brands

Official London Metal Exchange list of the brands deliverable in fulfilment of its ALUMINIUM ALLOY contract:

		Producers
Australia	AFFINITY	Affinity Metals Pty
Austria	AUSTRIA SEKUNDAR	Austria Sekundar Aluminium GmbH
	ALUMINIUM Ges. M.B.H. SAG	Salzburger Aluminium A/S
Belgium	ALUMET	Alumet
Canada	WABASH O	Wabash Alloys
Czech Republic	ALKOM	Kovohute Mnisek A/S
Denmark	GOTTHARD KOLDING	Gotthard Aluminium A/S
Finland	KUUSAKOSKI	Kuusakoski Oy
France	ALV	Aldevienne SA
	AP	Affinerie de Picardie
	AP-3	Refinal Industries SA
	NORD AFFINAGE	Nord Affinage SA
	SFRM	SFRM
	SNR.	Societe Nationale de Recuperation
Germany	BRUCH	Asperger Metallhutte Bruch GmbH
	EB	B.U.S. Berzelius Umwelt-Service AG
	KK	Karl Konzelmann GmbH
	OETINGER BERLIN	Metallwerk Oetinger Berlin Tempelhof GmbH
	UR-METALL	Metallhuttenwerke Bruch GmbH
	VAW	VAW Erftwerk
	WUPPERMETALL	Wuppermetall GmbH
India	HAMCO	Hindustan Alloys Mfg. Co. Ltd
Italy	GHIRAF	Eredi Di Ghidini Battista Srl
	PREMOLI	Premoli Luigi & Figli S.p.A.

	RIVA	Fonderie Riva S.p.A.
	R.M.	Raffmetal S.p.A.
	RMC	Raffineria Metalli Capra S.p.A.
	SACAL	Sacal Societa Alluminio Carisio S.p.A.
	VM	Vedani Carlo Metalli S.p.A.
Japan	AA	Altec Amino Co. Ltd
	AHRESTY	Ahresty Corporation
	ASK	Asahi Seiren Co. Ltd
	FAK	Fuji Alumi Industry Corporation
	HA	Hakata Aluminium Industrial Co. Ltd
	KAK	Kobayashi Aluminium Industrial Co. Ltd
	MCA1	M.C. Aluminium Co. Ltd
	MCA2	M.C. Aluminium Co. Ltd
	N	Naniwa Keikinzoku Co. Ltd
	S	Shikishima Aluminium Co. Ltd
	SANKO	Sanko Co. Ltd
	SZ	Goyo Kousan Co. Ltd
Poland	KETY A	Zaklady Metali Lekkich "Kety"SA
	KETY WR-35	Zaklady Metali Lekkich "Kety" SA
Spain	ALCASA	Aluminio Catalan SA 'Alcasa'
	INTAL	Industria Transformadora Del Aluminio SA
	REFINALSA	Refinados del Aluminio SA
	REMETAL	Remetal SA
Sweden	GOTTHARD ALMHULT	Gotthard Aluminium AB
Taiwan	NS	New Sun Metal Industry Co., Ltd
	SIGMA	Sigma Brothers Inc.
	TAKAO	Takao Aluminium Alloy Co. Ltd
Turkey	ALMET	Almet Aluminyum Sanayi ve Ticaret AS
U A E	SOLO UAE	Solo Industries Ltd
UK	A.C.	A. Cohen & Co. (Great Britain) Ltd
	ALCAN ENFIELD	Alcan Enfield Alloys Ltd
	ALDEC (SCOTLAND) LTD	Aldec (Scotland) Ltd
	ALDEC WESTBROM	Aldec Ltd
	ALENOY	Alcan Enfield Alloys Ltd
	AVON	Avon Metals Ltd
	BAT	B.A. Taylor (Metals) Ltd
	B.M.	Bernhard Metals (UK) Ltd
	BROCK	The Brock Metal Company
	BS	Brent Smelting Works Ltd
	COLESHILL ALUMINIUM	Coleshill Aluminium & Alloys Ltd
	C	Calder Aluminium Limited
	COOKSON *	Cookson Aluminium (Trent Alloys) Ltd.

**not good delivery if placed on warrant after 16.9.94*

	ESB	Presbar Diecastings Ltd
	H.H. WARDLE	H.H. Wardle (Metals) Ltd
	HLB FRY	H. Landseer Bailey Ltd
	MIL-VER	Mil-Ver Metal Co. Ltd
	MOTTAL	F.E. Mottram (Non-Ferrous) Ltd
	N.A.	Norton Aluminium Products Ltd
	NEESHAM	B.W. Neesham & Sons Ltd
	POWN-A L	J. McIntyre (Aluminium) Ltd
	RMM	Reg Morris (Brierley Hill) Ltd
	TAME VALLEY ALLOYS Ltd	Tame Valley Alloys Ltd
	WM1	IMI Wolverhampton Metal Ltd
	WM2	IMI Wolverhampton Metal Ltd
	WM3	IMI Wolverhampton Metal Ltd
USA	ALCHEM	Alchem Aluminium Inc.
	ALLIED CANAL	Allied Metal Co.
	ALLIED DIVISION	Allied Metal Co.
	BEHR	Behr Non-Ferrous Metals Inc.
	BERMCO	Bermco Aluminium
	CULP	Leggett & Platt Inc. (Culp Smelting)
	DELTA A	Delta Group Inc.
	LAB	Liston Aluminium Brick Co. of Corona
	LMI	Custom Alloy Sales Inc., Light Alloys Div.
	ROTH	Roth Bros. Smelting Corp.
	SAC	Spectro Alloys Corp.
	S.A.S	Mindis Consolidated Corp.
	SMI	State Metal Industries Inc.
	TIMCO	TST Inc.
	TMC	Thorock Metals Inc.
	TMP	Tower Metal Products L.P.
	USCO C	US Reduction Co.
	USCO EC	US Reduction Co.
	USCO M	US Reduction Co.
	USCO R	US Reduction Co.
	USCO T	US Reduction Co.
	VISTA	Vista Metals Corp.
	WABASH B	Wabash Alloys
	WABASH C	Wabash Alloys
	WABASH D	Wabash Alloys
	WABASH M	Wabash Alloys
	WABASH W	Wabash Alloys

Copper Brands

Official London Metal Exchange list of the brands deliverable in fulfilment of its COPPER - GRADE A contract:

CATHODES
Producers

Australia	ERANDS	Southern Copper Ltd
	ISA	Mount ISA Mines Ltd
	OLYDA	WMC (Olympic Dam Operations) Pty Ltd
Austria	BRX	Montanwerke Brixlegg GmbH
Belgium	SME	Metallo-Chemique International NV
	UMK	Union Miniere
Brazil	CbM	Caraiba Metals SA
Canada	CCR	Noranda Minerals Inc. - CCR Division
	FKA	Falconbridge Ltd
	ORC	Inco Ltd
Chile	AE	Corporacion Nacional del Cobre de Chile
	CCC	Corporacion Nacional del Cobre de Chile
	cCc-SX-EW	Corporacion Nacional del Cobre de Chile
	ENM	Empresa Nacional de Mineria
Finland	OKM	Outokumpu Mining
France	PAL	Cie Generale d'Electrolyse du Palais
Germany	HK	Huettenwerke Kayser AG
	KER	Huettenwerke Kayser AG
	NA-ES*	Norddeutsche Affinerie AG
	NA-ESN	Norddeutsche Affinerie AG
Italy	SAMETON	Enirisorse SpA
Japan	DOWA	Dowa Mining Co. Ltd
	HM	Nippon Mining Co. Ltd
	MITSUBISHI	Mitsubishi Metal Corp.
	MITSUI	Mitsui Mining & Smelting Co. Ltd
	OSR	Onahama Smelting & Refining Co. Ltd
	SR	Nippon Mining Co. Ltd

** not good delivery if placed on warrant after 10.6.94*

Japan	SUMIKO N	Sumitomo Metal Mining Co. Ltd
	SUMIKO T	Sumitomo Metal Mining Co. Ltd
	TAMANO	Mitsui Mining & Smelting Co. Ltd
Korea (South)	ONSAN	Lucky Metals Corp.
Norway	FHG	Falconbridge Nikkelverk AS
Oman	OMCO	Oman Mining Co. LLC
Peru	MP-ILO***	Southern Peru Copper Corporation
	SPCC-ILO ****	Southern Peru Copper Corporation
Philippines	PASAR	Philippine Associated Smelting
Poland	HMG-B	KGHM Polska Miedz SA
	HMG-S	KGHM Polska Miedz SA
	HML	KGHM Polska Miedz SA
South Africa	PMC	Palabora Mining Co. Ltd
Spain	ERCOSA	Elmet SL
	RTM	Rio Tinto Metal S.A.
Sweden	BOLIDEN	Boliden Ore & Metals
UK	JB	IMI Refiners Limited
USA	ATR	Asarco Inc.
	CB * CC	Cyprus Copper Co.
	KE*	Mitsubishi Metal America Corp.
	KUE	Kennecott Copper Corp.
	MAGMA	Magma Metals Co.
	MMB	Mitsubishi Metal America Corp.
	MSEW	Magma Metals Co.
	P * D	Phelps Dodge Sales Co. Inc.
	TB	Cyprus Copper Co.
Zambia	MCM	Zambia Consolidated Copper Mines Ltd
	REC	Zambia Consolidated Copper Mines Ltd

WIREBARS**

Zaire	UMK
Zambia	REC

** not good delivery if placed on warrant after 14.5.91*
*** not good delivery if placed on warrant after 31.12.93*
**** not good delivery if placed on warrant after 31.12.94*
***** good delivery if placed on warrant after 01.01.95*

Lead Brands

Official London Metal Exchange list of the brands deliverable in fulfilment of its STANDARD LEAD contract:

Argentina	NOAR
Australia	BHAS - BROKEN HILL AUSTRALIA (9997 on reverse)
	BHAS - BROKEN HILL AUSTRALIA (9999 on reverse)
Austria	BBU
Belgium	HOBOKEN EXTRA RAFFINE
	MC
	MCR MADE IN BELGIUM
Bulgaria	HP6
	KUM
	HP6
	OU3
Canada	AIM
	BM&S
	CANADA METAL
	NOVA PB
	TADANAC
	TONOLLI CANADA
China	IBIS
	MINER
	NH-R
	SKS
	TORCH
Denmark	BERA*
France	METALEUROP ME
	METALEUROP PYA
	SMMP
Germany	BHAS - WEST GERMANY 99985%

**not good delivery if placed on warrant after 19.7.94*

	BSB
	F
	ME OKER W 99.97
	ME WESER W 99.97
	NA (NORDDEUTSCHE AFFINERIE) E9999
	NA (NORDDEUTSCHE AFFINERIE) F99985
	NA (NORDDEUTSCHE AFFINERIE) H9997
	STOLBERG
India	INLEAD
Italy	ITALPIOMBO
	MAK-1
	MONTEVECCHIO
	S
	SAMIM
Japan	EMK-K
	EMK-T
	SK
	TAK
	THREE DIAMOND
Kazakhstan	C13
	YKCUK
Korea (North)	KM
Korea (South)	KZ-LEAD
Malaysia	MRISB MALAYSIA
Mexico	CMFyAM MEXICO
	IMM-CHIHUAHUA
	IMM-MONTERREY MEXICO
	MP-MP
	PENOLES
Macedonia	ZLETOVO 99.985%
Morocco	PZ MAROC
Myanmar	BM REFINED
Namibia	TCL
Peru	CP PERU INDUSTRIA PERUANA
Poland	H. 2OMS
Spain	FESA
	MTG
Sweden	BERA
	BOLIDEN A
	BOLIDEN B
UK	ASSOCIATED LEAD
	BLCO 9997%
	BLCO 9999%
	HJ ENTHOVEN & SONS
	LAA

	MELTON REFINED LEAD*
USA	DOE RUN
	EXIDE
	GLOVER
	GNB
	M
	OMAHA AND GRANT
	REVERE
	RMC
	RSR-CLFR
	RSR-INDY
	SANDERS
	SCHUYLKILL
	SEITZINGERS
former Yugoslavia	TREPCA

** not good delivery if placed on warrant after 19.7.94*

Nickel Producers

The production of the following producers shall constitute good delivery under the London Metal Exchange PRIMARY NICKEL contract provided it conforms with the terms and conditions of that contract.

	Locations
Bindura Nickel Corp. Ltd 70 Samora Machel Ave. Central Harare C4 Zimbabwe	Brand: TROJAN Bindura-Zimbabwe
Eramet SLN Tour Maine Montparnasse 33 Avenue du Maine 75755 Paris Cedex 15 France	Le Havre-France
Falconbridge Ltd Suite 1200 95 Wellington Street West Toronto, Ontario Canada M5J 2V4	Kristiansand-Norway
Impala Platinum Ltd Unicon House 70 Marshall Street Johannesburg 2001 South Africa	Springs-South Africa

INCO Ltd P.O.B. 44 Royal Trust Tower Toronto Dominion Centre Toronto, Ontario Canada M5K 1N4	Sudbury-Canada Thompson-Canada Clydach-UK
Jinchuan Non-Ferrous Metals Corporation Jinchang City Gansu Province The People's Republic of China	Brand: JINTUO GRADE 1
Marinduque Mining & Industrial Corp. Philippines	Nonoc Island-Philippines *Note: Marinduque Mining Industrial Corp. nickel is good for delivery only if placed on warrant before 28.6.85*
Norilsk Mining & Metallurgical Combine Nonferrous & Precious Metals Production Concern 'Norilskiy Nickel' Gvardec's Kaga Square, 2 663310 Norilsk Krasnoyarsk Region Russia	Norilsk-Russia
Outokumpu Harjavalta Metals Oy Harjavalta Works SF 29200 Harjavalta Finland	Harjavalta-Finland
Raznoimport** Moscow	Russia
Rio Tinto Mining (Zimbabwe) Ltd P.O.B. 8226 Causeway Harare Zimbabwe	Brand: EMPRESS Eiffel Flats-Zimbabwe

*** not good delivery if placed on warrant after 27.10.94*

Rustenburg Platinum Mines Ltd Consolidated Bldg Fox and Harrison Streets Johannesburg 2001 South Africa	Brand: MATTHEY NICKEL* (as produced after 1.4.80); Brand: RUSTENBURG NICKEL Rustenburg-South Africa
Severonickel Plant 7 Nikelevoe Shosse 184280 Monchegorsk Murmansk Region Russia	Brand: SEVERONICKEL
Sherritt Inc. 20 Queen Street West Suite 1012, Box 11 Toronto, Ontario Canada M5H 3R3	Ford Saskatchewan-Alberta Canada
Shimura Kako Co. Ltd Kokusai Building 1-1 Marunouchi 3-Chome Chiyoda-Ku Tokyo 100 Japan	Shimura-Japan
Sumitomo Metal Mining Co. Ltd 11-3, 5-Chome, Shimbashi Minato-Ku Tokyo Japan	Niihama-Japan
Companhia Niquel Tocantins Sao Miguel Paulista Sao Paulo - SP Brazil CEP 08090-000	Brand: TOCANTINS Sao Paulo-Brazil
Western Mining Corp. Ltd P.O.B. 80 Belmont W.A. 6104 Perth **Western Australia**	Kwinana-Australia

** not good delivery if placed on warrant after 3.9.93*

Tin Brands

Official London Metal Exchange list of the brands deliverable in fulfilment of its TIN contract:

Belgium	MC
Bolivia	ENAF
Brazil	AMAZON
	BEST
	CESBRA
	CORUMBA BRAZIL
	MAMORE
	MONSA HIGH GRADE TIN **
	TREVO
China	JH
	YT
Germany	M SPEZIAL **
Hong Kong	MAINLAND **
Indonesia	BANKA
	KOBA
	MENTOK
Malaysia	ESCOY LTD STRAITS REFINED TIN
	MALAYSIA SMELTING CORPORATION
Netherlands	WINDMILL I *
Nigeria	MAKERI
Singapore	ORCHID *
Thailand	THAISARCO
UK	FERRIBY HG *
USA	MIDCO

** not good delivery if placed on warrant after 7.1.94*
*** not good delivery if placed on warrant after 11.8.94*

Zinc Brands

SPECIAL HIGH GRADE ZINC

Official list of London Metal Exchange brands deliverable in fulfilment of contracts for SPECIAL HIGH GRADE ZINC:

Algeria	SNS SHG 99.99%+
Australia	AZ-SHG Zn 99995
Austria	BBU E-ZINK
Belgium	OVERCOR 99995**
	*** VM 99995+%
Brazil	CMM
	CPM ZINCO ELECTROLITICO
Bulgaria	KUM 99.995
Canada	CEZINC SHG
	HBMS CANADA SHG
	KIDD SHG
	COMINCO TADANAC MADE IN CANADA SHG
China	HX
	NH-SHG
	TORCH SHG
Finland	OUTOKUMPU ZINC
France	RCA
	V M *** 99995
	PENARROYA Z9
Germany	ME WESER E-ZINK
	MHD 99995+
	PREUSSAG E-ZINK*
	RMZ 99995%

** not good delivery if placed on warrant after 28.6.91*
*** not good delivery if placed on warrant after 10.5.94*

Italy	NUOVA SAMIM Zn 99.995
	PORTOVESME 99.995
	SHG P
Japan	AZC
	EMC-H
	EMC-K
	HSC-SHG
	MSC
	SK SHG
	THREE DIAMONDS SHG E.ZN
	TOHO EZ
Korea, (South)	KZ-SHG 99.995
	YP-SHG
Macedonia	ZLETOVO ZN 99.995
Mexico	IMM
	SLP
	PENOLES
Netherlands	AM&S Zn1
	BILLITON ZINK
Norway	NORZINK MADE IN NORWAY
	NORZINK ELEKTRO MADE IN NORWAY
Peru	MP LIMA SHG 99995
	CP PERU INDUSTRIA PERUANA 99995+
Spain	ASTUZINC ELECTRO 99.99+
	ASTUZINC ELECTRO 99.995%
	EDZ
Thailand	PADAENG THAILAND SHG
UK	BZL-Z1
USA	HV SHG ZN 99.995 min
	JMZ
Zaire	UZK 99995+

APPENDIX IV

Members

The London Metal Exchange Ltd

Chairman RK Bagri, CBE
Vice-Chairman R Kestenbaum
Chief Executive DE King

LME Managing Board – Representative Directors

GD Cohn
PCF Crowson
KC Davies
CJ Farrow
MJ Hutchinson
D Jackman
B Jones
LC Jones
MJG Linington
D Normark
B Östgren
JC Richards
GR Slater

RING DEALING MEMBERS - CATEGORY I

Company/Firm	*Address*
Amalgamated Metal Trading Ltd	Ground Floor Adelaide House London Bridge London EC4R 9DP Tel 0171 626 4521 Fax 0171 623 6015
Barclays Metals Ltd	6th Floor 2 Minster Court Mincing Lane London EC3R 7BB Tel 0171 283 6362 Fax 0171 621 5290
Billiton Metals Ltd	84 Fenchurch Street London EC3M 4BY Tel 0171 860 8250 Fax 0171 488 1986
Brandeis (Brokers) Ltd	Salters' Hall 4 Fore Street London EC2P 2NU Tel 0171 638 5877 Fax 0171 638 3031
Brody White (U.K.) Ltd	Devonshire House 146 Bishopsgate London EC2M 4JX Tel 0171 377 8050 Fax 0171 247 6470
Credit Lyonnais Rouse Ltd	Broadwalk House 5 Appold Street London EC2A 2DA Tel 0171 374 6100 Fax 0171 638 0886

Company/Firm	*Address*
Deutsche Sharps Pixley Metals Ltd	6-8 Bishopsgate London EC2P 2AT Tel 0171 971 7000 Fax 0171 971 7183
Gerald Ltd	Europe House World Trade Centre St. Katharine By The Tower London E1 9AA Tel 0171 867 9400 Fax 0171 867 9402
Lehman Brothers Commodities Ltd	One Broadgate London EC2M 7HA Tel 0171 601 0011 Fax 0171 260 2999
Metallgesellschaft Ltd	One America Square 17 Crosswall London EC3N 2LB Tel 0171 488 2244 Fax 0171 702 9650
Metdist Trading Ltd	80 Cannon Street London EC4N 6EJ Tel 0171 606 0463 Fax 0171 606 6650
Refco Overseas Ltd	Europe House World Trade Centre London E1 9AA Tel 0171 488 3232 Fax 0171 265 3959

Company/Firm	*Address*
Sogemin Metals Ltd	4th Floor 98 Cannon Street London EC4N 6EN Tel 0171 410 4950 Fax 0171 283 6788
Standard Chartered Bank	Mocatta House 4 Crosby Square London EC3A 6AQ Tel 0171 638 3636 Fax 0171 256 7750
Sucden (UK) Ltd	5 London Bridge Street London SE1 9SG Tel 0171 940 9400 Fax 0171 940 9500
Triland Metals Ltd	Bow Bells House Bread Street London EC4M 9BQ Tel 0171 236 5551 Fax 0171 489 1104
Rudolf Wolff & Co Ltd	D Section 2nd Floor Plantation House 31-35 Fenchurch Street London EC3M 3DX Tel 0171 626 8765 Fax 0171 626 3939

ASSOCIATE BROKER CLEARING MEMBERS - CATEGORY II

Company/Firm	*Address*
Aig International Ltd	120 Leman Street London E1 8EU Tel 0171 481 1098 Fax 0171 481 1633
Cargill Investor Services Ltd	Cannon Bridge, Level 2, 25 Dowgate Hill, London EC4R 2SL Tel 0171 283 5272 Fax 0171 929 4885
Chase Investment Bank Ltd	Woolgate House Coleman Street London EC2P 2HD Tel 0171 962 5213 Fax 0171 962 3019
Dean Witter International Ltd	1 Appold Street 6th Floor Broadgate 5 London EC2A 2AA Tel 0171 294 8500 Fax 0171 956 1247
Gni Ltd	4th Floor Atrium Building Cannon Bridge 25 Dowgate Hill London EC4R 2GN Tel 0171 337 3500 Fax 0171 337 3501
Goldman Sachs Futures	Peterborough Court 133 Fleet Street London EC4A 2BB Tel 0171 774 1000 Fax 0171 774 2020

Company/Firm	*Address*
Ing Derivatives	75 King William Street London EC4N 7EE Tel 0171 410 9944 Fax 0171 283 4321
Macquarie Securities (UK) Ltd	9th Floor Alban Gate 125 London Wall London EC2Y 5AS Tel 0171 776 7060 Fax 0171 776 7061
E.D. & F. Man International Ltd	Sugar Quay Lower Thames Street London EC3R 6DU Tel 0171 285 3000 Fax 0171 621 0149
MeesPierson ICS Ltd	Camomile Court 23 Camomile Street London EC3A 7PP Tel 0171 444 8000 Fax 0171 444 8888
Merrill Lynch Pierce Fenner & Smith (Brokers & Dealers) Ltd	Ropemaker Place 25 Ropemaker Street London EC2Y 9LY Tel 0171 628 1000 Fax 0171 867 2867
Mitsui Bussan Commodities Ltd	20 Old Bailey London EC4M 7QQ Tel 0171 248 6133 Fax 0171 236 0435
J.P. Morgan Securities Ltd	P.O.B. 161 60 Victoria Embankment London EC4Y 0JP Tel 0171 600 2300 Fax 0171 325 8267

Company/Firm	*Address*
Morgan Stanley & Co. International Ltd	25 Cabot Square Canary Wharf London E14 4QA Tel 0171 513 8000 Fax 0171 513 8990
PaineWebber International (U.K.) Ltd	1 Finsbury Avenue London EC2M 2PA Tel 0171 377 0055 Fax 0171 247 6039
Phibro Futures Ltd	Victoria Plaza 111 Buckingham Palace Road London SW1W 0SR Tel 0171 721 4555 Fax 0171 488 3116
Prudential-Bache (Futures) Ltd	9 Devonshire Square London EC2M 4HP Tel 0171 283 9166 Fax 0171 548 4433
N.M. Rothschild & Sons Ltd	New Court St. Swithin's Lane London EC4P 4DU. Tel 0171 280 5000 Fax 0171 280 5139
Standard Bank London Ltd	Cannon Bridge House 25 Dowgate Hill London EC4R 2SB Tel 0171 815 3000 Fax 0171 815 3099
UBS Futures & Options Ltd	100 Liverpool Street London EC2M 2RH Tel 0171 901 3333 Fax 0171 901 2345

ASSOCIATE TRADE CLEARING MEMBERS - CATEGORY III

Company/Firm	*Address*
BT Futures Corporation	130 Liberty Street New York NY 10006 U.S.A. Tel 001 212 775 3853 Fax 001 212 250 7499
Bank of America Illinois	162 Queen Victoria Street London EC4V 4BS Tel 0171 236 7444 Fax 0171 248 1244
Glencore UK Ltd	49 Wigmore Street London W1H 9LE Tel 0171 935 4455 Fax 0171 487 5431
Hunter Douglas N.V.	Gebouw Vierdrecht Ii Stationsweg 41 3331 Lr Zwijndrecht The Netherlands Tel 00 31 78 121888 Fax 00 31 78 123131
Hydro Aluminium A.S.	Drammensveien 264 P.O.Box 80 N-1321 Stabekk, Norway. Tel 00 47 22 73 81 00 Fax 00 47 22 73 78 18
Lazmet Ltd	Trans-World House 100 City Road London EC1Y 2BJ Tel 0171 972 0380 Fax 0171 972 0367
Toyota Tsusho U.K. Ltd	4th Floor 150 Holborn London EC1N 2NS Tel 0171 831 0903 Fax 0171 831 0776

ASSOCIATE BROKER MEMBERS - CATEGORY IV

Company/Firm	*Address*
J. Aron & Company (U.K.)	Peterborough Court 133 Fleet Street London EC4A 2BB Tel: 0171 774 1000 Fax: 0171 774 2020
Charles Davis (Metal Brokers) Ltd	One America Square 17 Crosswall London EC3N 2LB Tel: 0171 702 9735 Fax: 0171 702 9650
Dean Witter Reynolds Inc	2 World Trade Center 62nd Floor New York NY 10048 U.S.A. Tel: 001 212 392 2222 Fax: 001 212 392 4727
Lehman Brothers Inc	One Broadgate London EC2M 7HA Tel: 0171 601 0011 Fax: 0171 260 2999
Metchim Ltd	80 Cannon Street London EC4N 6EJ Tel 0171 606 0463 Fax 0171 606 6650
Smith Barney Europe Ltd	10 Piccadilly London W1V 9LA Tel: 0171 548 5555 Fax: 0171 548 5637

ASSOCIATE TRADE MEMBERS - CATEGORY V

Company/Firm	*Address*
A & M Minerals and Metals Ltd	140/142 St. John Street London EC1M 4AB Tel 0171 490 5574 Fax 0171 490 5294
Austmet UK Ltd	5 Old Queen Street London SW1H 9JA Tel 0171 799 1484 Fax 0171 222 0157
Ayrton & Partners Ltd	4 The Sanctuary Westminster London SW1P 3JS Tel 0171 222 2171 Fax 0171 222 5862
BHP Minerals Europe Ltd	29 Queen Anne's Gate London SW1H 9BU Tel 0171 222 1761 Fax 0171 222 0389
BICC Cables Ltd	P.O.Box 1 Prescot Merseyside L34 5SZ Tel 0151 430 8080 Fax 0151 430 2124
Thomas Bolton Ltd	P.O. Box 1 Froghall Stoke-On-Trent ST10 2HF Tel 01538 752241 Fax 01538 756715

Company/Firm	*Address*
Britannia Refined Metals Ltd	Botany Road Northfleet Gravesend Kent DA11 9BG Tel 01474 351188 Fax 01474 334013
Britannia Zinc Ltd	Kingsweston Lane Avonmouth Bristol BS11 8HT Tel 0117 982 3646 Fax 0117 982 5568
Calder Industrial Materials Ltd	Crescent House Newcastle-Upon-Tyne NE99 1GE Tel 0191 261 0161 Fax 0191 261 1001
Chile Copper Ltd	York House 23 Kingsway London WC2B 6UJ Tel 0171 240 0221 Fax 0171 240 8031
Cominco Ltd	1500 - 120 Adelaide St. W., Toronto Ontario Canada M5H 1T1 Tel 001 416 8691850 Fax 001 416 8628553
Enirisorse Spa	Piazza Lodovica Cerva, 7 00143 Rome Italy Tel 00 396 59841 Fax 00 396 54646033

Company/Firm	*Address*
Euromin S.A.	Bowater House East 68 Knightsbridge London SW1A 7LT Tel 0171 917 8960 Fax 0171 917 8962
Europa Metalli-LMI Spa	Borgo Pinti 99 50121 Florence Italy Tel 00 39 55 49741 Fax 00 39 55 2344492
Freeport Mcmoran Copper & Gold Inc.	1615 Poydras Street New Orleans LA 70112, U.S.A. Tel 001 504 582 4913 Fax 001 504 582 1835
W.J. Furse & Co. Ltd	Wilford Road Nottingham NG2 1EB Tel 0115 986 3471 Fax 0115 986 0538
Gecamines Commerciale	Building Gecamines Bld Du 30 Juin, B.P. 13998 Kinshasa Republique du Zaire Tel 0171 321 0355 (London Office) Fax 0171 321 0787
Halcor Metal Works S.A.	115 Kifissias Avenue Athens 115 24 Greece Tel 00 30 1 69 20 137 Fax 00 30 1 69 17 927

Company/Firm	*Address*
"Hirsch" Kupfer-Und Messingwerke GmbH	Postfach 200521 41205 Monchengladbach Germany Tel 00 49 21 66927930 Fax 00 49 21 66272692
Huttenwerke Kayser AG	Postfach 1560 D-44505 Lunen Germany Tel 00 49 23 061080 Fax 00 49 23 108449
IMI Plc	c/o Imi Refiners Ltd James Bridge Copper Works Darlaston Road Walsall WS2 9SJ Tel 01922 721292 Fax 01922 25218
Inco Ltd	P.O. Box 44 Royal Trust Tower Toronto-Dominion Centre Toronto Ontario Canada M5K 1N4 Tel 001 416 3617511 Fax 001 416 3617781
Indumetal Recycling S.A.	Kauka Erreka Bidea, 1 48150 Asua - Sondika (Vizcaya) Spain Tel 00 3444 530800 Fax 00 3444 530120
The International Metals & Minerals Company Ltd	7th Floor 20 Old Bailey London EC4M 7EP Tel 0171 248 6134 Fax 0171 236 0497

Company/Firm	*Address*
KM-Rohstoff GmbH	Klosterstrasse 29 D-49074 Osnabruck Germany Tel 00 49 541 321 4900 Fax 00 49 541 321 4930
Lazarus Metals Ltd	Exchange Tower 1 Harbour Exchange Square London E14 9GE Tel 0171 971 5640 Fax 0171 971 5649
Lewis & Peat (Metals) Ltd	21 Knightsbridge London SW1X 7LY Tel 0171 235 0099 Fax 0171 623 3351
Lonconex Ltd	No 1, Warwick Row London SW1E 5ER Tel 0171 834 6767 Fax 0171 630 0575
Memaco Services Ltd	Memaco House 215 Marsh Wall London E14 9FJ Tel 0171 712 3800 Fax 0171 678 8321
Memaco (UK) Ltd	Memaco House 215 Marsh Wall London E14 9FJ Tel 0171 895 9977 Fax 0171 895 9797

Company/Firm	*Address*
Metal Distributors (UK) Ltd	45 Circus Road St.John's Wood London NW8 9JW Tel 0171 266 2939 Fax 0171 266 2310
Metdist Ltd	80 Cannon Street London EC4N 6EJ Tel 0171 606 0463 Fax 0171 606 6650
Minemet S.A.	8c Avenue de Champel Case Postale 316 1211 Geneva 12 Switzerland Tel 00 41 22 8394220 Fax 00 41 22 3478816
Mitsui & Co UK Plc	20 Old Bailey London Tel 0171 822 0464 Fax 0171 236 2130
Moscow Narodny Bank Ltd	81 King William Street London EC4P 4JS Tel 0171 623 2066 Fax 0171 283 4840
NKT A/S	NKT Alle 1 DK-2605 Brondby Denmark Tel 00 45 43 48 20 10 Fax 00 45 43 96 03 02

Company/Firm	*Address*
Noranda Sales Corporation of Canada Ltd	20 Bedfordbury Covent Garden London WC2N 4TP Tel 0171 497 2046 Fax 0171 497 2748
Normaco Ltd	1 Jermyn Street London SW1Y 4UH Tel 0171 222 3999 Fax 0171 839 1111
Norddeutsche Affinerie AG	Hovestrasse 50 D-20539 Hamburg Germany Tel 00 49 40 78830 Fax 00 49 40 78832255
Outokumpu Rawmet (Switzerland) Ltd	Lagerhausstrasse 3 Postfach 665 Ch-8401 Winterthur Switzerland Tel 00 52 212 1721 Fax 00 52 212 1720
Pasminco Europe Ltd	P.O.Box 237 4th Floor, Hanover House Queen Charlotte Street Bristol BS99 7EA Tel 01179 304234 Fax 01179 297307
Pirelli General Plc	P.O.Box 6 Leigh Road Eastleigh Hampshire SO50 9YE Tel 01703 644522 Fax 01703 332754

Company/Firm	*Address*
Primary Industries (UK) Ltd	Carrier House 1-9 Warwick Row London SW1E 5ER Tel 0171 834 6767 Fax 0171 630 0575
RMT Metals Ltd	33/34 Chiswell Street London EC1Y 4SE Tel 0171 638 4789 Fax 0171 382 9708
RTZ Ltd	6 St.James's Square London SW1Y 4LD Tel 0171 930 2399 Fax 0171 930 3249
Sherritt Inc	Sales & Marketing, Sherritt Metals 10101 - 114 Street Fort Saskatchewan Alberta Canada T8L 2P2 Tel 001 403 998 6911 Fax 001 403 998 6568
Steeple Industries Ltd	12th Floor One America Square 17 Crosswall London EC3N 2LB Tel 0171 265 5493 Fax 0171 265 5492
The Metal & Commodity Company Ltd	One America Square 17 Crosswall London EC3N 2LB Tel 0171 488 2288 Fax 0171 480 5913

Company/Firm	*Address*
Tower Commodities Ltd	c/o Meespierson Ics Ltd Camomile Court 23 Camomile Street London EC3A 7PP Tel 0171 444 8000 Fax 0171 444 8888
Toyota Tsusho Corporation	Toyota Building 7-23 Meieki 4-Chome Nakamura-Ku Nagoya 450 Japan Tel 0171 831 0833 (London Office) Fax 0171 831 0776
Toyota Tsusho (Hong Kong) Co., Ltd	27th Floor, United Centre No.95, Queensway Hong Kong Tel 00 852 5294361 Fax 00 852 5295839
Wilhelm Grillo Handelsgesellschaft MBH	Am Grillopark 5 47169 Duisburg Germany Tel 00 49 20340660 Fax 00 49 203404302
WMC Nickel Sales Corporation	1 First Canadian Place, Suite 970 P.O.Box 76 Toronto Ontario, Canada M5X 1B1 Tel 001 416 366 0132 Fax 001 416 366 6444

APPENDIX V

Warehouses

Warehouses listed here for the storage of LME accepted metal are not bound to accept consignments from every source even though such metal may be included on an LME brand list. Local regulations should be checked before the metal is shipped.

Neither The London Metal Exchange Limited nor any of its Directors or other officers shall be under any liability whatsoever either in contract or in tort to any Member or other person in respect of any act or omission in relation to the listing of any warehouse or the maintenance of any such listing.

Abbreviations used in this appendix for metals that may be stored:

Al- primary aluminium	Ni- nickel
AA- aluminium alloy	Sn- tin
Cu- copper	Zn- zinc
Pb- lead	

LONDON METAL EXCHANGE LISTED WAREHOUSES

Companies	Depots	Metals stored
BELGIUM		
ANTWERP		
Antwerp Combined Terminals NV Rostockweg Kade 312 2030 Antwerp 3	Warehouse De Schelde Quay No. 308 Rostockweg	Al AA Cu Pb Ni Sn Zn
	Warehouse ACT 472 Churchilldok 2030 Antwerp	Al Cu Pb Ni Sn Zn
Metal Terminals International NV Van Meterenkaai 1 2000 Antwerp	Warehouse West 1 Land Van Waasland 9130 Beveren - Kallo	Al Cu Pb Ni Sn Zn
	Warehouse West 5 Land Van Waasland 9130 Beveren - Kallo	Al AA Cu Pb Ni Sn Zn
	Compounds West 1/X, 3, 5/X, 6, 21 & 28 Land Van Waasland 9130 Beveren - Kallo	Al

Companies	Depots	Metals stored
Noord Natie NV Stadswaag 7-8 B2000 Antwerp	Warehouse 243 Quay 243, 4th Harbourdock	Al Cu Pb Sn Zn
	Warehouse 320 Quay 320, 6th Harbourdock	Al Cu Pb Zn
	Warehouse 2 Quay 702, 9th Harbourdock	Al Cu Pb Ni Sn Zn
Ocean Group (Belgium) NV McGregor Cory Division Antwerpsebaan T.O. Kaai 730 DD B2040 Antwerp	Blauwgaren Kaai 734	Al Cu Pb Zn
	Compound Denmark Vrasendok, Kade No. 1227 DD	Al
Rhenus Belgium NV Noorderlaan 139 2030 Antwerp	Warehouse no. 4 Nietweg	Al Cu Pb Ni Sn Zn
C. Steinweg NV Haven 73 Ouland 3 B2030 Antwerp	Albert I Albert Dock 115, Vrieskaai 2030 Antwerp	Al Cu Pb Zn
	Albert II Albert Dock 113, Vrieskaai 2030 Antwerp	Al Cu Pb Zn
	Albert III Albert Dock 131, Vrieskaai 2030 Antwerp	Al Cu Pb Zn
	Albert VI Albert Dock 133, Vrieskaai 2030 Antwerp	Al Cu Pb Zn
	Baltimore Albert Dock, Vrieskaai 2030 Antwerp	Al Cu Pb Zn
	Warehouse Bernard/Alfons Albert Dock 125/7, Vrieskaai 2030 Antwerp	Al Cu Pb Zn
	Warehouse Bombay 1/2 Rijstweg Albert Dock 2030 Antwerp	Al Cu Pb Zn
	Warehouse Brisbane Albert Dock 117, Vrieskaai 2030 Antwerp	Al Cu Pb Zn
	Fortuna Albert Dock 121, Vrieskaai 2030 Antwerp	Al Cu Pb Zn
	Kemzeke Hulsterstraat Noord 1 9190 Kemzeke-Stekene	AA
	Warehouse Ketelhof Ketelhofstraat 2030 Antwerp	Al Cu Pb Ni Sn Zn
	Warehouse Mercier Schenkeldijk 2030 Antwerp	Al Cu Pb Zn
	Missouri Schenkeldijk 2030 Antwerp	Al Cu Pb Ni Sn Zn

Companies	Depots	Metals stored
	Oosterweel II Wilmardonksteenweg 2030 Antwerp	Al Cu Pb Sn Zn
	Ouland Haven 73 - Ouland 3	Al Cu Pb Ni Sn Zn
	Warehouse Parsifal Nieuwlandenweg 24 2030 Antwerp	Al Cu Pb Zn
	Veurne Albert Dock 119, Vrieskaai 2030 Antwerp	Al Cu Pb Zn
	Warehouse Waasland Hasopweg 5 Vrasendok 1199 - Kallo, Beveren	Al Cu Pb Zn
	Warehouse 125/127 Albert Dock, Vrieskaai 2030 Antwerp	Al Cu Pb Zn
	Warehouse 742 Delwaidedok Nieuwe Westweg 2040 Antwerp	Al Cu Pb Zn
	Compound 742/A Delwaidedok Nieuwe Westweg 2040 Antwerp	Al
	Compound Albert IV Albert Dock 131/3, Vrieskaai 2030 Antwerp	Al
	Compound Oosterweel Wilmarsdonksteenweg 2030 Antwerp	Al
	Compound Ouland Ouland	Al
Zuidnatie NV Oude Leeuwenrui 48 B2000 Antwerp	Warehouse Iris Klein Zuidland 9 2030 Antwerp	Al Cu Pb Ni Sn Zn
	Warehouse 144 Mulhouselaan Zuid 3/5 2030 Antwerp	Al Cu Pb Ni Sn Zn

Companies	Depots	Metals stored
FRANCE		
DUNKIRK		
Pinnacle Storage Ltd (Hull) c/o Societe Francaise de Transports Gondrand Freres 11/19 Quai du Risban BP 3513 59383 Dunkerque Cedex	Warehouse 3, Rue Vancauwenberghe Zone Industrielle de Petite-Synthe 59640 Dunkerque	Al AA Cu Pb Ni Sn Zn
C. Steinweg-Handelsveem BV (Rotterdam) c/o Societe Generale de Transit Consignation, Magasinage et Affretement - SOGETRA 10 Quai de la Citadelle B.P. 3126 59377 Dunkerque-Cedex 1	Warehouse 1 Rue Flipo Prolongee 59210 Coudekerque Branche	Al Cu Pb Ni Sn Zn
GERMANY		
BREMEN		
Bremer Lagerhaus-Gesellschaft AG Hafenstrasse 55 28217 Bremen	Warehouse 13 Freeport /Ueberseehafen 28217 Bremen	Al Cu Pb Sn Zn
	Warehouse 16 Freeport /Ueberseehafen 28217 Bremen	Al Cu Pb Ni Sn Zn
Steinweg Handelsveem GmbH (Kap Horn Strasse) Postfach 103647 28036 Bremen	Warehouse 4, Kalihafen 28237 Bremen	Al Cu Pb Sn Zn
DUISBURG		
C. Steinweg GmbH (Koenigstrasse 13) Postfach 101302 47013 Duisburg	DFL & DFL Lager 2 Im Freihafen 4 47138 Duisburg	AA
HAMBURG		
Eurokai KGaA Griesenwerder Damm 6 21129 Hamburg	Warehouses 1, 2, 3, 5, 6 & 8 Zellmannstrasse 21129 Hamburg	Al Cu Pb Ni Sn Zn
	Warehouse 5, Einsiedeldeich 1-3 20539 Hamburg	Al Cu Pb Ni Sn Zn
	Warehouses D & N, Rossweg 20 20457 Hamburg	Al Cu Pb Ni Sn Zn
Gerd Buss Lager-und Speditions-Gesellschaft Kuhwerder m.b.h. Nehlsstrasse 55 – 20457 Hamburg	Warehouse 73, Nehlsstrasse 20457 Hamburg	Al AA Cu Pb Ni Sn Zn

Companies	Depots	Metals stored
C. Steinweg-Handelsveem GmbH Steinhoft 5-7 Haus am Fleet 20459 Hamburg	Warehouses 60 & 63, am Kamerunkai 20457 Hamburg	Al Cu Pb Ni Sn Zn
	Warehouse 64 (compartments A & C) Kamerunweg 20457 Hamburg	Al Cu Pb Ni Sn Zn
	Warehouses 65-I & 65-II, Worthdamm 20457 Hamburg	Al Cu Pb Ni Sn Zn
	Warehouses 93 I, II, III, V, VI, VII, VIII, IX, XI & XII, Pollhornweg 21107 Hamburg	Al Cu Pb Ni Sn Zn
Berthold Vollers G.m.b.h. Brandenburgerstrasse 3-5 20457 Hamburg	Warehouse Spreehafen 1-15 (compartments 1-15) Brandenburgerstrasse 3-5 20457 Hamburg	Al Cu Pb Sn Zn
	Warehouse Spreehafen 1-15 (compartments 4-15) Brandenburgerstrasse 3-5 20457 Hamburg	AA
	Warehouse Spreehafen 1-15 (compartments 1 & 6) Brandenburgerstrasse 3-5 20457 Hamburg	Ni

GREECE

PIRAEUS

Companies	Depots	Metals stored
Almatrans S.A 81 Notara Street 18510 Piraeus	Warehouse No. 1 Nato Avenue Aspropygos	Al Cu Pb Sn Zn
Sarmed Hellenic Warehouse. Sarantitis S.A 81 Akti Maiouli 18538 Piraeus	Warehouse No. 3 Old National Road Athens - Thebes Mandra Attica (Elefsis) 196.00	Al Cu Pb Ni Sn Zn

ITALY

BRESCIA

Companies	Depots	Metals stored
B. Pacorini S.r.l. Punto Franco Nuovo 34123 Trieste	Warehouse No. 80 Via S. Giovanni Bosco 17 25127 Brescia 1	AA

GENOA

Companies	Depots	Metals stored
Genoa Metal Terminal S.r.l. Via Conservatori del Mare 5/12 16123 Genoa	Warehouses 1, 2, 3 & 4 Etiopia Pier	Al Cu Pb Ni Sn Zn
	Libia Lana Libia Pier Genoa	Al Cu Pb Zn

Companies	Depots	Metals stored
B. Pacorini S.r.l. Via Sampierdarena 12/5 16149 Genoa	Warehouse No. 27 Via Rivarolo 2A Rivarolo 16161	Al AA Cu Pb Ni Sn Zn
	Warehouse No. 30 Via Rivarolo 2A Rivarolo 16159	Al AA Cu Pb Ni Sn Zn

LEGHORN

Companies	Depots	Metals stored
F. lli Bartoli Via S. Orlando 36/38 (P.O.B. 172) 57123 Leghorn	Warehouse 69/J Attracco No. 54 Calata Pisa Porto Commerciale	Al Cu Pb Ni Sn Zn
	Warehouse 211E Via S. Orlando 38 Leghorn	Al Cu Pb Ni Sn Zn

TRIESTE

Companies	Depots	Metals stored
McGregor Cory Cargo Services S.r.l. Viale Miramare No.5 34135 Trieste	Warehouses 8 & 16 Punto Franco Vecchio 34100 Trieste	Al AA Cu Pb Zn
	Warehouse 27 Punto Franco Vecchio 34100 Trieste	Al AA Cu Pb Ni Sn Zn
	Warehouse 30 Punto Franco Vecchio 34100 Trieste	Al Cu Pb Zn
	Warehouse 49 Punto Franco Nuovo 34123 Trieste	Al AA Cu Pb Ni Sn Zn
	Warehouses 55, 58 & 61 Punto Franco Nuovo 34123 Trieste	Al Cu Pb Ni Sn Zn
B. Pacorini S.r.l. Punto Franco Nuovo 34123 Trieste	Warehouse 4 (sections 3, 9 & 10) Punto Franco Vecchio 34100 Trieste	Pb Zn
	Warehouse 5 Punto Franco Vecchio 34100 Trieste	Al AA Cu Pb Zn
	Warehouse 9 Punto Franco Vecchio 34100 Trieste	Al Pb Zn
	Warehouse 28 Punto Franco Vecchio 34100 Trieste	Al Cu Pb Zn
	Warehouse 36 & Timber Shed Scalo Legnami 34100 Trieste	Al Cu Pb Zn
	Warehouse No. 42 Scalo Legnami 34100 Trieste	Al AA Cu Pb Zn
	Warehouse 50 Punto Franco Nuovo	Al Cu Pb Zn

Companies	Depots	Metals stored
	Warehouse 53 Punto Franco Nuovo	Al AA Cu Pb Ni Sn Zn
	Warehouse 57 (compartment 7/8) Punto Franco Nuovo	Al AA Cu Pb Ni Sn Zn
	Warehouse 57 (compartments 20, 514, 524, 534, 544 & 554) Punto Franco Nuovo	AA
	Warehouse 57B (compartment 1) Punto Franco Nuovo	Al AA Cu Pb Sn Zn
	Warehouse 59 Punto Franco Nuovo	Al Cu Pb Zn
	Warehouses 60 (compartments 7, 204, 304, 404, 504 & 604), 73B & 73C Punto Franco Nuovo	Al AA Cu Pb Ni Sn Zn
	Warehouse 70 Punto Franco Nuovo	Al Cu Pb Ni Sn Zn
	Warehouse 72 (Section 3) Punto Franco Nuovo	Al AA Cu Pb Zn
	Warehouse 73A Punto Franco Nuovo	Al AA Cu Pb Ni Sn Zn
	Compounds 4, 5 & 6 Scalo Legnami 34100 Trieste	Al
	Compounds 38 & 49 Scalo Legnami 34100 Trieste	Al

JAPAN

HAKATA

Companies	Depots	Metals stored
Sumitomo Warehouse Co. Ltd c/o Hakata Rinko Soko Co. Ltd 11-1 Okihama-Cho Hakata-Ku Fukuoka City	Hakata Rinko Soko Hakozaki Pier Terminal 8-6 Hakozaki-Futo 1-Chome Higashi-Ku	Al

KOBE

Companies	Depots	Metals stored
Mitsubishi Warehouse & Transportation Co. Ltd 19-1 Nihonbashi 1-Chome Chuo-Ku Tokyo 103	DaiNiShinko No. H 6-22 Onohama-Cho Chuo-Ku Hyogo Prefecture	Al
	Rokko No. D Nishi 2-Chome, Koyo-Cho Higashinada-Ku Hyogo Prefecture	Al
Mitsui Soko Co. Ltd 13-12 Kayabacho 1-Chome Nihonbashi, Chuo-Ku Tokyo 103	Onohama 5-11 Onohama-Cho Chuo-Ku	Al

Companies	Depots	Metals stored
Sumitomo Warehouse Co. Ltd Tokyo Sumitomo Twin Building East 23rd Floor 27-1 Shinkawa 2-Chome Chuo-Ku Tokyo 104	Port Island Terminal 14 Minatojima, 7-Chome Chuo-Ku Kobe 650	Al
MOJI		
Mitsubishi Warehouse & Transportation Co. Ltd 19-1, Nihonbashi 1-Chome Chuo-Ku Tokyo 103	Tachinoura Terminal No. A 20-8 Tachinoura-Kaigan Moji-Ku Kitakyushu Fukuoka Prefecture	Al
Mitsui Soko Co. Ltd 13-12 Kayabacho 1-Chome Nihonbashi, Chuo-Ku Tokyo 103	Tachinoura 12-1 Tachinoura-Kaigan Moji-Ku Kitakyushu 801	Al
NAGOYA		
Japan Transcity Corporation Chitose-cho 6-6 Yokkaichi 510	Kasumi YK Warehouse 2-1 Kasumi Yokkaichi City, Mie Prefecture	Al
	Nagoya-Nishi-4-Ku Warehouse 1-5-7 Higashihama Tobishima-Mura Ama-Gun, Aichi Prefecture	Al
Meiko Trans Co. Ltd Meiko Building 4-6 Irifune 2-Chome, Minato-Ku Nagoya 455	2-15-2 Higashihama Tobishima-Mura, Ama-Gun Aichi Prefecture	Al
Mitsubishi Warehouse & Transportation Co. Ltd 19-1 Nihonbashi 1-Chome, Chuo-Ku Tokyo 103	Nishi 4-Ku No. A 23-3 Higashihama 2-Chome Tobishima-Mura, Ama-Gun Aichi Prefecture	Al
	Ohe No. A 7 Ohe-Cho, Minato-Ku Aichi Prefecture	Al
Mitsui Soko Co. Ltd 13-12 Kayabacho 1-Chome Nihonbashi, Chuo-Ku Tokyo 103	Chikko 2-1-37 Chidori, Minato-Ku	Al
	Nishi 4-Ku 2-23-2 Higashihama Tobishima-Mura Ama-Gun Aichi Prefecture 490-14	Al
Steinweg-Suzue (Japan) Co. Ltd c/o Suzue Corporation No. 14-24 Kaigan 1-Chome Minato-Ku Tokyo 105	Warehouse A2 Nishi 4 2-1-10 Higashihama Tobishima-Mura Ama-Gun Aichi Prefecture	Al

Companies	Depots	Metals stored
Sumitomo Warehouse Co. Ltd Tokyo Sumitomo Twin Building East 23rd Floor 27-1 Shinkawa 2-Chome Chuo-Ku Tokyo 104	Nagoya Port Terminal Warehouse No. 100 1-12 Higashihama 2-Chome Tobishima-Mura, Ama-Gun, Aichi Prefecture	Al
	Nakagawa Terminal Warehouses No. 100 and 200 1 Tamafune-Cho 3-Chome Nakagawa-Ku Nagoya 452	Al
Toyo Warehouse Co. Ltd 2-10 1-Chome Irifune, Minato-Ku Nagoya 455	Seibu Warehouses No. 24 1-9 H & 25 5-7 G Higashihama 2-Chome Tobishima-Mura Ama-Gun, Aichi Prefecture	Al
Yasuda Warehouse Co. Ltd International Forwarding Department 3-3-8 Kaigan Minato-Ku Tokyo 108	Nagoya No. 6 Tokai Warehouse Centre 5-7 Higashima 1-Chome Tobishima-Mura Ama-Gun, Aichi Prefecture	Al

OSAKA

Companies	Depots	Metals stored
Mitsubishi Warehouse & Transportation Co. Ltd 19-1 Nihonbashi 1-Chome Chuo-Ku Tokyo 103	Nanko R 2-46 Nanko-Kita 3-Chome Suminoe-Ku	Al
	Sakurajima Warehouses D, E & E6 4-11 Sakurajima 3-Chome Konohana-Ku	Al
Shibusawa Warehouse Co. Ltd 13-16 1-Chome Kayabacho, Nihonbashi Chuo-Ku, Tokyo 103	Chikko Warehouse 4-5-4 Kaigan Dori Minato-Ku	Al
Sumitomo Warehouse Co. Ltd Sumitomo Twin Building East 23rd Floor 27-1 Shinkawa 2-Chome Chuo-Ku Tokyo 104	Nangan Terminal Warehouses No. 414 & 415 6-15 Kaigandori 2-Chome Minato-Ku Osaka 552	Al
	Nanko Terminal 5-3 Nanko Naka 6-Chome Suminoe-Ku Osaka 559	Al
Yasuda Warehouse Co. Ltd International Forwarding Department 3-3-8 Kaigan Minato-Ku Tokyo 108	Fukusaki 1-148 Fukusaki 3-Chome Minato-Ku	Al

Companies	Depots	Metals stored
YOKOHAMA		
Yamatane Corporation 15-4 Daikoku-Futo Tsurumi-Ku Yokohama 230	Yamatane Daikoku-Futo 15-4 Daikoku-Futo Tsurumi-Ku Yokohama 230	Al
Mitsubishi Warehouse & Transportation Co. Ltd 19-1 Nihonbashi 1-Chome, Chuo-Ku Tokyo 103	Daikoku No. A Warehouse 6 Daikoku-Futo Tsurumi-Ku Yokohama Kanagawa Prefecture	Al
	Daikoku No. B Warehouse 15 Daikoku-Futo, Tsurumi-Ku, Yokohama Kanagawa Prefecture	Al
Mitsui Soko Co. Ltd 13-12 Kayabacho 1-Chome Nihonbashi, Chuo-Ku Tokyo 103	Daikoku 15 Daikoku-Futo Tsurumi-Ku	Al
	Honmoku 1 Honmoku-Futo Naka-Ku	Al
Nissin Corporation No. 84 6-Chome Onoe-Cho, Naka-Ku Yokohama 231	Nissin East Warehouse 3 Honmoku-Futo, Naka-Ku	Al
	Nissin Kanagawa Warehouse No. 9 Ebisucho Kanagawa-Ku	Al
Shibusawa Warehouse Co. Ltd 13-16 1-Chome Kayabacho, Nihonbashi Chuo-Ku, Tokyo 103	Yamashita Futo Warehouse 279-1 Yamashita-Cho Naka-Ku	Al
Steinweg-Suzue (Japan) Co. Ltd c/o Suzue Corporation No. 14-24 Kaigan 1-Chome Minato-Ku Tokyo 105	Daikoku L-1 5-1 Daikoku-Futo Tsurumi-Ku	Al
	Daikoku-Pier Warehouse Daikoku-Futo 1 Tsurumi-Ku	Al
	Honmoku-Pier C Berth Warehouse Honmoku Futo-1 Naka-Ku	Al
	Yamashita No. 12 Yamashita-Cho Naka-Ku	Al
Sumitomo Warehouse Co. Ltd Tokyo Sumitomo Twin Building East 23rd Floor 27-1 Shinkawa 2-Chome Chuo-Ku Tokyo 104	Daikoku Pier Terminal 7-2 Daikoku-Futo, Tsurumi-Ku Yokohama 230	Al
	Daikoku Terminal Warehouse No. 200 15 Daikoku-Futo, Tsurumi-Ku Yokohama 230	Al

Companies	Depots	Metals stored
Sumitomo Warehouse Co. Ltd	Honmoku No. 1 Pier Terminal 7 Shin-Yamashita, 3-Chome, Naka-Ku Yokohama 231	Al
	Honmoku No. 2 Pier Terminal 1 Honmoku-Futo, Naka-Ku Yokohama 231	Al
Tokyo InternationalTerminal Inc. Room 802 Shinbashiekimae Building 20-15 Sinbashi 2-Chome, Minato-Ku Tokyo 105	No.4 Suzushige-Cho Kanagawa-Ku	Al
Yasuda Warehouse Co. Ltd International Forwarding Department 3-3-8 Kaigan Minato-Ku Tokyo 108	Daikoku & Daikoku YS-I I5 Daikoku-Futo, Tsurumi-Ku	Al
	Honmoku 'B' Honmoku-Futo Naka-Ku	Al

NETHERLANDS

AMSTERDAM

Hoogewerff De Rijke Terminals BV Sardinieweg 4 1044 AE Amsterdam	Warehouse No. 3 Valreep 13 Haven West 1042 AN	Al Cu Pb Ni Sn Zn
	Warehouse 11 & 12 Vlothavenweg 10 1013 BJ	Al Cu Pb Ni Sn Zn
	Warehouse Hornhaven Corsicaweg 10 1044 AB	Al Cu Pb Ni Sn Zn
C. Steinweg-Handelsveem BV Sardinieweg 7 1044 AG Amsterdam	Warehouse Handelsveem 1 Sardinieweg 7 1044 AG	Al Cu Pb Ni Sn Zn
	Warehouse Handelsveem 3 Sextantweg 8 1042 AH	Al Cu Pb Ni Sn Zn
Verenigd Cargadoorskantoor BV (VCK) (Valreep 13) P.O.B. 561 1000 AN Amsterdam	Warehouse No. 6 Valreep 13 1044 AN	Al Cu Pb Ni Sn Zn

MOERDIJK

Metaal Transport BV (Heyplaatweg 16) P.O.B. 54500 3008 KA Rotterdam/Heyplaat	Warehouses No. 3 & No. 4 Valsweg 19 4782 PW Moerdijk	Al Cu Pb Ni Sn Zn

Companies	Depots	Metals stored
ROTTERDAM		
Albatros Shipping & Transport BV (Eemhavenweg 125) P.O.B. 54052 3008 JB Rotterdam	Warehouse Albatros Eemhavenweg 125 3089 KE	Al Cu Pb Ni Sn Zn
	Warehouse Laysan Eemhavenweg 125 3089 KE	Al AA Cu Pb Ni Sn Zn
	Compound Antarctica Eemhavenweg 125 3089 KE	Al
Hollands Veem BV (Boyneweg 1) P.O.B. 1337 3180 AH Rozenburg	Hollands Veem I Marconistraat 52 3029 AK	Al Cu Pb Sn Zn
	Hollands Veem II Schiehaven 15 3024 EC	Al Cu Pb Sn Zn
Hoogewerff De Rijke Terminals BV (Welplaatweg 26) P.O.B. 5145 3197 ZG Botlek Rotterdam	Warehouse 1 Bunschotenweg 141 3089 KB	Al Cu Pb Ni Sn Zn
	Warehouse Nos. 1 & 2 Distripark Nieuwesluisweg 224 3197 KV Botlek	Al Cu Pb Ni Sn Zn
	Warehouses A & B Welplaatweg 26 3197 KS Botlek	Al Cu Pb Ni Sn Zn
	Compound 1X Bunschotenweg 141 3089 KB	Al
	Compound C Welplaatweg 26 3197 KS Botlek	Al
	Distripark, Distripark II & Distripark III Compounds Nieuwesluisweg 224 3197 KV Botlek	Al
MetaalTransport BV (Heyplaatweg 16) P.O.B. 54500 3008 KA Rotterdam/Heyplaat	Warehouse Gevelco Heyplaatweg 4 3089 JC	Al Cu Pb Ni Sn Zn
	Warehouse Metra Heyplaatweg 16 3089 JC	Al Cu Pb Ni Sn Zn
	Warehouse Meyers Heyplaatweg 10 3089 JC	Al Cu Pb Ni Sn Zn
	Compound Gevelco Heylaatweg 4 3089 JC	Al
	Compound Metra Heyplaatweg 16 3089 JC	Al
	Compound Meyers Heyplaatweg 10 3089 JC	Al

Companies	Depots	Metals stored
Multi-Terminals Waalhaven BV (50 Pannerdenstraat) P.O.B. 5242 3008 AE Rotterdam	Hector 1 & 2 Waalhaven Pier 2 3008 AE	Al Cu Pb Sn Zn
Ocean Warehousing BV (41 Basaltstraat) P.O.B. 5955 3008 AZ Rotterdam	RHB Warehouses C & D, Waalhaven 3087 BL	Al Cu Pb Zn
	Warehouse A Nieuwesluisweg 100 3197 KV Botlek	Al Cu Pb Ni Sn Zn
	Warehouse G (RHB) Ophemerstraat 67 3089 JD	Pb Zn
	Warehouse H Willingestraat 33 3087A	Cu Pb Ni Zn
	Warehouse I 1e Eemhavenstraat 7 3089 KD	Cu Pb Ni Zn
	Warehouse 35 Mijestraat 35 3044 CE	Al Pb Zn
	Compound A Nieuwesluisweg 100 3197 KV Botlek	Al
J. Oosterom & Zoon Stuwadors en Controlebedrijf BV Rijnhaven Z.Z.7 5-7 3072 AJ Rotterdam	Warehouse Oost Rijnhaven Z.Z.7 5-7 3072 AJ	Al Cu Pb Ni Sn Zn
Roehlig & Co.'s Expeditie en Scheepvaartbedrijf BV (Glashaven 54) P.O. B. 901 3000 AX Rotterdam	Kofschip Bunschotenweg 141B 3089 KB	Al Cu Pb Ni Sn Zn
	RHB Waalhaven N.Z. 4 & O.Z. 22 3087 BL	Al Cu Pb Ni Sn Zn
C. Steinweg-Handelsveem BV (Parmentierplein 1) P.O.B. 1068 3000 BB Rotterdam	Apollo 1/2 Pannerdenstraat 50 Waalhaven	Al Cu Pb Zn
	Argentinie & Brazilie Keilestraat Lekhaven	Al Cu Pb Ni Sn Zn
	Argentinie D1, D2 & India D Theemsweg 26 Seinehaven	Al Cu Pb Zn
	Baltimore I, II, III & Norfolk II Kramer Pier Prof. Gerbrandyweg Botlek	Al Cu Pb Ni Sn Zn
	Beatrix 1, 2, 3, 18 and 19 Spakenburgweg 45 Beatrixhaven	Al Cu Pb Ni Sn Zn
	Beatrix I, III & IV Spakenburgweg 45 Beatrixhaven	Al Cu Pb Ni Sn Zn

Companies	Depots	Metals stored
	Beatrix II Spakenburgweg 45 Beatrixhaven	Al Cu Pb Zn
	Celebes, Java, Borneo & Bali Lloydstraat Sint Jobshaven	Al Cu Pb Zn
	China (Japan), China II (Korea), China III (Siam) & China IV (Eersteling) Parallelweg Rijnhaven	Al Cu Pb Ni Sn Zn
	Europort A, B, & C Ir de Grootweg Spijkenisse	Al Cu Pb Zn
	Fenix I & II Veerlaan Rijnhaven	Al Cu Pb Ni Sn Zn
	Fenix West Rijnhaven 22 Rijnhaven	Al Cu Pb Ni Sn Zn
	Hercules 1/6 (compartments 3/6) Pannerdenstraat 50 Waalhaven	Al Cu Pb Ni Sn Zn
	Hercules 1/6 (compartments 1/2) Pannerdenstraat 50 Waalhaven	Al Cu Pb Zn
	India, Levant, Pakistan & Uruguay Keilestraat Lekhaven	Al Cu Pb Zn
	India P: L, M & N Elbeweg 101 Dintelhaven	Al Cu Pb Ni Sn Zn
	Jobshaven S: CB Schiehavenweg Schiehaven	Al Cu Pb Zn
	MarcoNil Radiostraat, Merwehaven	Al Cu Pb Ni Sn Zn
	MarcoNilII: A, B, C & D Gustoweg Merwehaven	Al Cu Pb Zn
	Norfolk Wilhelminakade Rijnhaven	Al Cu Pb Zn
	Norfolk I Zuidzijde Markweg Dintelhaven, Europort 5	Al Cu Pb Ni Sn Zn
	Roos C Nijmegenstraat 25-27 Waalhaven	Al Cu Pb Ni Sn Zn
	Warehouse Roos G3 (F & H), GM 1 & 2 Parmentierplein 1	Al Cu Pb Ni Sn Zn
	Santos Brede Hilledijk Rijnhaven	Al Cu Pb Zn

Companies	Depots	Metals stored
	Shannon I Shannonweg 10 Botlek	Al Cu Pb Ni Sn Zn
	Sluisjedijk I & II Sluisjedijk Waalhaven	Al Cu Pb Ni Sn Zn
	Waalhaven Roos 25A & 25B Pier II Nijmegenstraat Waalhaven	Al Cu Pb Ni Sn Zn
	Waalhaven Roos F1 & 2, G1 & 2, H1, 2, 3 & 4 & S1, 2 & 3 Pier 1 Dodewaardstraat Waalhaven	Al Cu Pb Ni Sn Zn
	Compound Apollo Pannerdenstraat 50 Waalhaven	Al
	Compound Argentinie Keilestraat Lekhaven	Al
	Compound Argentinie D Theemsweg 26 Seinehaven	Al
	Compound Beatrix Spakenburgwerg 45 Beatrixhaven	Al
	Compound Botlek Kramer Pier Prof. Gebrandyweg Botlek	Al
	Compound Botlek II Shannonweg 10 Botlek	Al
	Compound Celebes Lloydstraat Sint Jobshaven	Al
	Compound China Parallelweg Rijnhaven	Al
	Compound Dintelhaven Zuidzijde Markweg Dintelhaven, Europoort 5	Al
	Compound Europoort Ir de Grootweg 18 Spijkenisse	Al
	Compound Fenix Veerlaan Rijnhaven	Al
	Compound Fenix II Rijnhaven 22 Rijnhaven	Al
	Compound India G Botlekweg/Montrealweg	Al
	Compounds India PI & II Elbeweg 101 Dinterhaven	Al

Companies	Depots	Metals stored
	Compound Jobshaven S: C Schiehavenweg Schiehaven	Al
	Compound MarcoNilll Gustoweg Merwehaven	Al
	Compound Roos C Nijmegenstraat 25-27 Waalhaven	Al
	Compound Roos GM Parmentierplein 1	Al
	Compound Roos 25 Pier 11, Nijmegenstraat Waalhaven	Al
	Compound Sluisjedijk Sluisjedijk Waalhaven	Al
	Compounds Waalhaven Roos F, G H & S Pier 1 Dodewaardstraat Waalhaven	Al

VLISSINGEN

Companies	Depots	Metals stored
N.V. Haven van Vlissingen Stadhuisplein 20 P.O.B. 398 4380 AJ Vlissingen.	Alleghany VI Engelandweg Quay Nr 1054 Vlissingen - Oost	Cu Pb Ni Sn Zn

SINGAPORE

Companies	Depots	Metals stored
Cornelder Hoogewerff (Singapore) Pte Ltd No. 1 Tuas Avenue 3 Singapore 2263	Warehouse 3 No. l Tuas Avenue 3 Singapore 2263	Al Cu Pb Ni Zn
Mac-Nels Agencies Pte Ltd Block 1 No.01-01/03-01 Pasir Panjang Districentre Pasir Panjang Road Singapore 0511	Block 1 01-02 Pasir Panjang Districentre Pasir Panjang Road Singapore 0511	Al Cu Pb Ni Zn
	Warehouse No. PB1 Pasir Panjang Terminal PSA gate 5 Pasir Panjang Road Singapore 0511	Al AA Cu Pb Ni Zn
McGregor Cory Pte Ltd No. 12 Tanjong Penjuru Crescent Singapore 2260	Warehouse No. 1 29 Pioneer Sector 1 Singapore 2262	Al Cu Pb Zn
Scandinavia Warehouses Pte (Ltd) 34 Penjuru Lane Singapore 2260	Warehouse No. 1 34 Penjuru Lane	Al Cu Pb Ni Zn

Companies	Depots	Metals stored
C. Steinweg Warehousing (FE) Pte Ltd Jurong Port Administration Building 37 Jurong Port Road Singapore 2261	J1, J2, J9 & J11 Jurong Port 37 Jurong Port Road Singapore 2261	Al Cu Pb Ni Zn
	J14, J14A & J14B Jurong Port 37 Jurong Port Road Singapore 2261	Al Cu Pb Zn
	PB-4, PB-5, PT 10 & PT 11 Pasir Panjang 15 West Coast Highway Singapore 0511	Al Cu Pb Zn
	Compounds J1, J14A & J14B Jurong Port 37 Jurong Port Road Singapore 2261	Al

SPAIN

BARCELONA

Barcelona Metals SA Muelle Fontente s/n Puerto de Barcelona 08039 Barcelona	Warehouse No. 1 Tinglado 5 (Sects. 7,8,9) Muelle de Poniente 08039 Barcelona	Al AA Cu Pb Ni Sn Zn
Henry Bath & Son Ltd Carretera Bilbao-Plencia 48150 Asua-Sondika Vizcaya	Tinglado 5 (Sects. 1,2,3) Muelle de Poniente 80839 Barcelona	Al Cu Pb Ni Sn Zn
Saport-Mapor SA Muelle Adosado Sur Puerto de Barcelona 08039 Barcelona	Muelle Adosado Sur s/n Puerto de Barcelona 08039 Barcelona	Al AA Cu Pb Ni Sn Zn

BILBAO

Henry Bath & Son Ltd Carretera Bilbao-Plencia 48150 Asua-Sondika Vizcaya	Warehouse No. 1 Muelle Reina Victoria 48980 Santurce	Al Cu Pb Ni Sn Zn
	Warehouse 3, Ground Floor CanAlde Deusto No. II 48014 Deusto	Al Cu Pb Ni Sn Zn
	Warehouse No. 5 Zona Portuaria 48980 Santurce	Al Cu Pb Ni Sn Zn
	Compound No. 4 CanAlde Deusto Entrado de San Felicisimo Deposito Franco de Bilbao 48014 Deusto	Al

Companies	Depots	Metals stored
SWEDEN		
GOTHENBURG		
AB Skandiatransport (Packhusplatsen 2) S-403 43 Gothenburg	Warehouse D Free Harbour	Al AA Cu Pb Ni Sn Zn
	Warehouse E Free Harbour	Al Cu Pb Zn
	Warehouses E & F Salsmastaregatan 25	Al Cu Pb Ni Sn Zn
	Warehouse No.3 Gotaverken Gothenburg Harbour	Al Cu Pb Ni Sn Zn
	Warehouses 107, 113 & H Free Harbour	Al Cu Pb Ni Sn Zn
HELSINGBORG		
AB Skandiatransport (Packhusplatsen 2) S-403 43 Gothenburg	Gamla Castor Sjogatan	Al Cu Pb Ni Sn Zn
	Warehouse Gipsladan Helsingborg Harbour	Al Cu Pb Zn
	Hagen Helsingborg Harbour	Al Cu Pb Ni Sn Zn
	Oresunds 1 Helsingborg Harbour	Al Cu Pb Zn
	Oresund 2 South Harbour	Al Cu Pb Zn
	Oresund 3 South Harbour	Al Cu Pb Zn
	Scansped 1 & 2	Al Cu Pb Zn
	Warehouse Stora Castor Sjogatan	Al Cu Pb Ni Sn Zn
	Warehouse 404 Oceanterminalen	Al Cu Pb Zn
	Warehouses 405, 407 & 408 Oceanterminalen	Al Cu Pb Zn
	Warehouse 600 Helsingborg Harbour	Al Cu Pb Ni Sn Zn
	Warehouse 705 S-Terminal Helsingborg Harbour	Al Cu Pb Ni Sn Zn
	Warehouse 906 Vasthamnen	Al Cu Pb Ni Sn Zn
	Compounds South Harbourg 1, 2 & 3 S-Terminal Helsingborg Harbour	Al
	Compound Plan Castor Sjogatan	Al
	Compound Plan Oresund Helsingborg Harbour Sydhamnsgaten	Al
	Compound Plan 1a, 1b & 1c South Harbour	Al

Companies	Depots	Metals stored

UNITED KINGDOM

AVONMOUTH

Companies	Depots	Metals stored
Henry Bath & Son Ltd Acornfield Road Knowsley Industrial Park North Kirkby, Liverpool L33 7UG	Warehouse T & Warehouse T North St Andrew's Road Avonmouth BS11 9DQ	Al Cu Pb Zn
	Compound T St Andrew's Road Avenmouth BS11 9DQ	Al
	Portbury & Portbury No. 2 Compounds RoyAlPortbury Dock Port of Bristol	Al
McGregor Cory (UK) Severn Road Avonmouth Bristol BS11 0YQ	Site 1, Sections 4,5 & 6 Severn Road Chittening Avonmouth	Al Cu Pb Sn Zn
	Site 2, Warehouse A Avonmouth Docks Estate Chittening Avonmouth BS11 0YB	Al Cu Pb Sn Zn

DERBY

Companies	Depots	Metals stored
Pinnacle Storage Ltd Cat and Fiddle Lane West Hallam Derby DE7 6HE	Warehouse 4954 & 6772 Cat and Fiddle Lane	AA

GOOLE

Companies	Depots	Metals stored
Pinnacle Storage Ltd West Carr Lane, Stoneferry Hull HU7 0BS	Knedlington Road Howden near Goole DN14 7EG	Al Cu Pb Zn

HULL

Companies	Depots	Metals stored
Henry Bath & Son Ltd Acornfield Road Knowsley Industrial Park North Kirkby, L33 7UG	Warehouse Unit 4 Eastfield Road South Humber Road South Killingholme Immingham DN40 3DU	Al Cu Pb Ni Sn Zn
	Warehouse Unit 7 Laporte Road Stallingborough Immingham	Al Cu Pb Zn
	Compound Unit 4 Eastfield Road South Humber Road South Killingholme Immingham DN40 3DU	Al

Companies	Depots	Metals stored
Limb Stevedores (Hull) Ltd Alexandra Dock Terminal Alexandra Dock Hull HU9 1TA	Warehouse No. 1 King George Dock Hull	Al Cu Pb Ni Sn Zn
Pinnacle Storage Ltd West Carr Lane, Stoneferry Hull HU7 OBS	Riverside (compartments 11 & 26) Foster Street Stoneferry Industrial Park Hull	Al Cu Pb Ni Sn Zn
	Riverside (compartment 17) Foster Street Stoneferry Industrial Park Hull	Al
	Warehouse 16 Stoneferry Wharves Hull	Al AA Cu Pb Ni Sn Zn
	Compounds A & B West Carr Lane Stoneferry	Al

LIVERPOOL

Companies	Depots	Metals stored
Henry Bath & Son Ltd Acornfield Road Knowsley Industrial Park North Kirkby, Liverpool L33 7UG	South Hornby 1, 2, 3 & 4, & North Alexandra Liverpool Freeport	Al Cu Pb Ni Sn Zn
	Warehouse Huskisson Sylo Regent Road Liverpool L20	Al Cu Pb Ni Sn Zn
	Kirkby 1 & 2 Acornfield Road Knowsley Industrial Park North Kirkby	Al AA Cu Pb Ni Sn Zn
	Kirkby 4 Acornfield Road Knowsley Industrial Park North Kirkby	Al AA Cu Pb Ni Sn Zn
	Compounds North Alexandra No. 1 & No. 2 Liverpool Freeport	Al
	Compound Regent Regent Road Liverpool Freeport	Al
	Kirkby Compounds East, North & 4 Acornfield Road Knowsley Industrial Park North Kirkby	Al
	Strand Entrance Compound Liverpool Freeport	Al

Companies	Depots	Metals stored
Henry Diaper & Co. Ltd P.O. Box 25 Storage House, Lees Road Knowsley Industrial Park North Liverpool L33 7SD	Warehouses 1, 2, 10, 11 & 12 Lees Road Knowsley Industrial Park North (Kirkby Sector) Liverpool L33 7SD	Al Cu Pb Sn Zn
	Warehouses 9 & 13 Lees Road Knowsley Industrial Park North (Kirkby Sector) Liverpool L33 7SD	Al AA Cu Pb Sn Zn
Pinnacle Storage Ltd Great Howard Street Liverpool L3 7DL	Warehouses A & E Effingham Street Bootle	Al AA Cu Pb Ni Sn Zn
Warrant Distribution Limited Warrant House 18 Rumford Court Rumford Place Liverpool L3 9DG	Warehouses No. 2, No. 8 Unit 5 & No. 8 Unit 6 Lees Road Knowsley Industrial Park North Kirkby, Liverpool L33 7SD	Al Cu Pb Ni Sn Zn
	Warehouse No. 3 N1 Huskisson Dock Regent Road Bootle Liverpool L3 0AT	Al Cu Pb Ni Sn Zn
	Warehouse No. 4 Blackstone Street Bootle Liverpool L5 9TB	Al Cu Pb Ni Sn Zn
	Warehouse No. 5 Heysham Road Aintree Liverpool L30 6UR	Al Cu Pb Ni Sn Zn
	Warehouse No. 7 Lees Road Knowsley Industrial Park North Kirkby, Liverpool L33 7SD	Al Cu Pb Zn
	Warehouse No. 23 Field Lane	Al Cu Pb Ni Sn Zn
	Compound North 1 N1 Huskisson Dock Regent Road Bootle Liverpool L3 0AT	Al
	Compound North 2 & 3 East Waterloo Dock Regent Road Bootle Liverpool L3 0AT	Al
	Compound N4 Seaforth Terminal Liverpool	Al
	Compound N5 Heysham Road Aintree Liverpool L30 6UR	Al

Companies	Depots	Metals stored
NEWCASTLE		
Warrant Distribution Limited Warrant House 18 Rumford Court Rumford Place Liverpool L3 9DG	Warehouse A Riverside Quay Port of Tyne	Al Cu Pb Ni Sn Zn
	Compounds A1 & A2 Riverside Quay Port of Tyne	Al
SUNDERLAND		
North European Marine Services Ltd South Dock Sunderland Tyne & Wear SR1 2BU	Warehouse Endurance South Dock Sunderland Tyne & Wear SR1 2BU	Al Cu Pb Ni Sn Zn
	Compounds Arrow & Endurance South Dock Sunderland Tyne & Wear	Al

UNITED STATES OF AMERICA

Companies	Depots	Metals stored
BALTIMORE		
Hoogewerff (USA) Inc. 2807 West Busch Blvd Tampa Florida 33618	6905 San Tomas Road Elkridge Maryland 21227	Al Pb Ni Sn Zn
C. Steinweg (Baltimore) Inc. Baltimore Freeport Centre 2500 Broening Highway Baltimore Maryland 21224	B.F.C. 1 2500C Broening Highway	Al Pb Ni Sn Zn
	B.F.C. 2 2400D Broening Highway	Al Pb Ni Sn Zn
	B.F.C. 5 100 Vail Street Baltimore Maryland 21224	Al Pb Ni Sn Zn
	Broening Compound Broening Highway	Al
BRIDGEPORT		
Connecticut Metals Warehouse Inc. 914 Arctic Street Bridgeport Connecticut 06608	Warehouses 723 & 876 939 Barnum Avenue	Al Pb Ni Sn Zn
CHICAGO		
S.H. Bell Co. 1378 Freeport Road P.O. Box 11495 Pittsburgh Pennsylvania 15238-3190	Warehouse No. 1 Port of Indiana Burns International Harbor 6600 U.S. Highway 12 Portage Indiana 46368	Al Cu Pb Ni Sn Zn

Companies	Depots	Metals stored
Hoogewerff Reserve Ltd Partnership 12701 South Doty Avenue Chicago Illinois 60633	Warehouse 2 12800 S. Butler Drive Illinois 60633	Al Cu Pb Ni Sn Zn
	Back-up Warehouse 12800 S. Butler Drive Illinois 60633	Al AA Pb Ni Sn Zn

DETROIT

Hoogewerff Motor City. Warehouses Inc 4005 West Fort Street Detroit Michigan 48209	4005 West Fort Street Detroit	Al AA Pb Ni Sn Zn
Metro International Trade Services Inc. 6850 Middlebelt Road Romulus Michigan 48174	6490 Lynch Road Detroit	Al AA Pb Ni Sn Zn
	17423 West Jefferson Riverview Michigan 48192	Al Pb Ni Sn Zn
	Warehouse No. 1 NTD Storage 360 Great Lakes Avenue Ecorse, Detroit Michigan 48229	Al Pb Ni Sn Zn
	Riverview Compound 17423 West Jefferson Riverview Michigan 48192	Al

LOS ANGELES/LONG BEACH

Metro International Trade Services Inc. 6850 Middlebelt Road Romulus Michigan 48174	4552 Brickell Privado Ontario California	Al AA Pb Ni Sn Zn
	Warehouse A 1500 West Dominguez Street Long Beach California 90810	Al AA Pb Ni Sn Zn

NEW HAVEN

New Haven Terminal Inc. 100 Waterfront Street P.O. Box 9423 New Haven Connecticut 06534-0423	Warehouses 9 & 10 100 Waterfront Street	Al Pb Ni Sn Zn

PITTSBURG

S.H. Bell Co. 1378 Freeport Road P.O. Box 11495 Pittsburgh PA 15238	Warehouse 4/5 One Third Street Braddock Pennsylvania 15104	Al Pb Ni Sn Zn

Companies	Depots	Metals stored
ST LOUIS		
The Delivery Network Inc. Highway 3 & Bissell P.O.Box C Madison Illinois 62060	1100 Niedringhaus Granite City Illinois 62040	Al Cu Ni Sn Zn
	2000 Access Boulevard Madison Illinois 62060	Al AA Cu Ni Sn Zn
	Compound 2000A Access Boulevard Madison Illinois 62060	Al
	Compound 1100 Niedringhaus Granite City Illinois 62040	Al
TOLEDO		
Toledo Metals Terminals Inc. 3332 St Lawrence Drive Toledo Ohio 43605	Warehouse 3332 St Lawrence Drive	Al Pb Ni Sn Zn
	Compound 3332 St Lawrence Drive	Al

APPENDIX VI

Samplers and assayers

Organisations listed in this appendix have, as required, confirmed their competence to follow the standard procedures laid down by the London Metal Exchange.

Alex Stewart (Assayers) Ltd
Caddick Rd
Knowsley Industrial Estate
Knowsley
Merseyside
L34 9ER

Tel. 0151-548-7777
Telex 627759
Fax 0151-548-0714

AMTAC Laboratories Ltd
Norman Road
Broadheath
Altrincham
Cheshire
WA14 4EP

Tel. 0161-928-8924
Telex 665337
Fax 0161-927-7359

Inspectorate Griffith Ltd
Perry Rd
Witham
Essex
CM8 3TU

Tel. 01376-515081
Telex 995281
Fax 01376-520819

Johnson Matthey PLC
(Materials Technology Division)
Orchard Rd
Royston
Herts
SG8 5HE

Tel. 01763-253000
Telex 818375
Fax 01763-2445014

Kitto Laboratories
Christy House
Church Lane
Bocking
Braintree
Essex
CM7 5RX

Tel. 01376-552554
Telex 98166
Fax 01376-552923

Alfred H. Knight International Ltd
Eccleston Grange
Prescot Rd
St Helens
Merseyside
WA10 3BQ

Tel. 01744-33757
Telex 629648
Fax 01744-27062

Rooney Laboratories Ltd
7 Lennox Mall
Lister Road
Basingstoke
Hampshire
RG22 4DF

Tel. 01256-469941
Fax 01256-470669

SGS United Kingdom Ltd
(Minerals Division)
SGS House
John's Lane
Tividale
Warley
W. Midlands
B69 3HX

Tel. 0121-520-6454
Telex 339118
Fax 0121-520-2438

Tioxide Analytical Services
West Site
Haverton Hill Road
Billingham
Cleveland
TS23 1PS

Tel. 01642-376767
Telex 58351
Fax 01642-561853

Alex E Tucker
(Birmingham Central Laboratories)
20 Key Hill Drive
Hockley
Birmingham
B18 5NY

Tel. 0121-554-2353
Fax 0121-554-2353

APPENDIX VII

Market calls

FIRST SESSION

Aluminium alloy	11.45 to 11.50
Tin	11.50 to 11.55
Primary aluminium	11.55 to 12.00
Copper	12.00 to 12.05
Lead	12.05 to 12.10
Zinc	12.10 to 12.15
Nickel	12.15 to 12.20
Interval	12.20 to 12.30

Official prices

Copper	12.30 to 12.35
Interval	12.3 5 to 12.40
Tin	12.40 to 12.45
Lead	12.45 to 12.50
Zinc	12.50 to 12.55
Primary aluminium	12.55 to 13.00
Nickel	13.00 to 13.05
Aluminium alloy	13.05 to 13.10

Kerb trading follows declaration of Official Prices until 13.30

SECOND SESSION

Lead	15.20 to 15.25
Zinc	15.25 to 15.30
Copper	15.30 to 15.35
Primary aluminium	15.35 to 15.40
Tin	15.40 to 15.45
Nickel	15.45 to 15.50
Aluminium alloy	15.50 to 15.55
Interval	15.55 to 16.00
Lead	16.00 to 16.05
Zinc	16.05 to 16.10
Copper	16.10 to 16.15
Primary aluminium	16.15 to 16.20
Tin	16.20 to 16.25
Nickel	16.25 to 16.30
Aluminium alloy	16.30 to 16.35

Kerb trading follows 16.35-17.00 with nickel phasing out at 16.45, zinc and lead at 16.50, aluminium alloy and tin at 16.55 and copper and primary aluminium at 17.00.

APPENDIX VIII

Requirements for the listing of brands of metals

---

1. An application for listing a brand of metal must be submitted on behalf of the producer through a Ring Dealing Member who is responsible for ensuring the accuracy and completeness of the application and the reliability and competence of the applicant producer.

2. Guidance notes for listing of brands of each metal are available from Ring Dealing Members or the secretariat of the Exchange. Applicants should not proceed with an application prior to reading the applicable guidance notes.

   The general procedures outlined in the guidance notes include the following requirements:-

   2.1 That the producer supplies a written undertaking that the brand to be listed will conform to the quality of the applicable Exchange contract and that such quality will be maintained in accordance with the lots supplied for testing purposes (see 2.2 below). Also, the producer will undertake to investigate any complaints as to the quality of the brand of metal without time limit.

   2.2 That testing of commercial quantities of the metal be undertaken by Exchange approved consumers who will provide the Exchange with information as to metal quality, suitability and compliance with the applicable Exchange Contract.

2.3 That, except in the case of copper and aluminium alloy, the application for listing, once testing is completed, is to be accompanied by certificates of analysis applicable to the test lots from two assayers accredited by the Exchange

2.4 That a fee is paid to the Exchange for the listing of a brand.

3. Subsequent to listing any alteration or additions to the details given at the time of listing by the producer are to be notified to the Exchange.

4. Brands are listed at the discretion of the Directors, who are entitled to make such investigations into the applicant as they deem appropriate, and no listing shall become valid until particulars of the brand concerned have been posted on the Exchange for 28 days.

5. Should any complaint as to the quality of any listed brand be brought to the notice of the Directors, they shall have the right to investigate the brand in question and the right to suspend deliveries onto warrant until quality is proven.

6. The Directors may, at their discretion, waive or amend any of the listing requirements.

7. Neither The London Metal Exchange Limited nor any of its Directors or other officers shall be under any liability whatsoever either in contract or in tort to any Member or other person in respect of any act or omission in relation to the listing of any brand of metal or the maintenance of any such listing.